Food Chemistry in Small Bites

FOOD CHEMISTRY IN SMALL BITES

Patricia B. O'Hara

Molecular and scientific illustrations by Richard A. Blatchly

UNIVERSITY OF CALIFORNIA PRESS

University of California Press
Oakland, California

Library of Congress Cataloging-in-Publication Data

Names: O'Hara, Patricia (Patricia B.), author. | Blatchly, Richard, illustrator.
Title: Food chemistry in small bites / Patricia B. O'Hara ; molecular and scientific illustrations by Richard A. Blatchly.
Description: Oakland, California : University of California Press, [2025] | Includes bibliographical references and index.
Identifiers: LCCN 2024028967 (print) | LCCN 2024028968 (ebook) | ISBN 9780520397620 (cloth) | ISBN 9780520397637 (paperback) | ISBN 9780520397644 (ebook)
Subjects: LCSH: Food—Analysis.
Classification: LCC TX541 .O395 2025 (print) | LCC TX541 (ebook) | DDC 664/.07—dc23/eng/20240910
LC record available at https://lccn.loc.gov/2024028967
LC ebook record available at https://lccn.loc.gov/2024028968

GPSR Authorized Representative: Easy Access System Europe, Mustamäe tee 50, 10621 Tallinn, Estonia, gpsr.requests@easproject.com

34 33 32 31 30 29 28 27 26 25
10 9 8 7 6 5 4 3 2 1

publication supported by a grant from
The Community Foundation for Greater New Haven
as part of the *Urban Haven Project*

To my girls, Sarah and Becca,
to my feline guardian angel, Lola,
and to the memory of my beloved Rich

Contents

Figures

Tables

Boxes

Acknowledgments

I credit two terrific students, Xiao Xiao and Vivian Mac, both of the class of 2016 at Amherst College, for convincing me that food chemistry belongs in the chemistry curriculum. The passion of these dedicated students was transformative. Out of that study grew a lab course for nonscience majors. Most recently, over one hundred students signed up to take Molecular Gastronomy and Food Science. It is a testament to the power of student-driven initiatives and how they can shape educational experiences.

The cooperation of Amherst College's Dining Services has been essential to the successful running of this lab-based cooking course. Two outstanding directors, Charlie Thompson and Joe Flueckiger, believed in me and the course and supported it with food supplies and expert guidance from their staff. Each spring, they made possible our offering of a food festival for the community.

In 2018, we graduated from cooking in kitchenettes scattered around campus to a dedicated culinary teaching lab. The current room is set up as a swing space and can be converted from a lab to a gathering space with minimum effort. Jim Brassord, Director of Facilities Planning and Operation, made this possible.

The Harvard University course Science and Cooking: From Haute Cuisine to Soft Matter Science, taught by Michael Brenner, David Wetz, and Pia Sörensen, was a keystone for me as I was developing my course and this book.

As the book took shape, feedback from high school teachers, especially Roger Wallace, was essential. Students in the class also provided feedback. Caroline Stole and Brandon Kwon, both of the class of 2023, helped with images. Donna Nestor, class of 2024, was the force that got me to the finish line.

From the moment the proposal for *Food Chemistry in Small Bites* landed on the desk of Chloe Layman, acquisitions editor at the University of California Press, she supported, advised, and mentored me as the work transitioned from proposal to manuscript to this textbook.

Finally, the book would not have happened without the constant encouragement of my beloved husband, Richard Blatchly, who passed away in the very final stages of the book's production. Rich created some of the graphs and every molecular image in the book. He submitted the molecules to the molecular modeling and computational program Spartan to generate their unique surface charge models. We promised each other a life full of surprises over forty years ago. *Food Chemistry in Small Bites* is the final and most ambitious addition to that list.

Introduction

Food Chemistry in Small Bites breaks the science of cooking into easy-to-understand pieces. It's perfect for someone just starting to learn about science or for anyone who wants to know more about what happens to their food as they cook.

The book is divided into four parts.

Informing (Chapters 1 and 2): This part gives you the background information you need. It covers the basics of chemistry that will help you understand the science of cooking.

Transforming (Chapters 3–5): This part explores how cooking changes our food. It delves into chemical and biochemical concepts like combustion (burning), reduction and oxidation (redox) reactions, acid and base reactions, isomerization, and molecular structure and recognition.

Reforming (Chapter 6): The focus here is on how we can change our cooking habits to create less food waste, which contributes to greenhouse gases and climate change. At the same time, it explains that we can make our meals healthier by switching to a plant-based diet.

Performing (Chapter 7): The final part presents experiments that will help you put some of the conceptual ideas into practice. It brings together all the knowledge from the previous parts and shows how it applies to creating delicious and satisfying meals.

The goal of *Food Chemistry in Small Bites* is to connect everyday cooking processes with fundamental scientific concepts. So whether you're boiling water, frying an egg, or making yogurt, this book helps you understand the chemistry behind it in a simple and engaging way.

PART I

INFORMING

When we step into a new country, it helps us fully enjoy our trip if we have familiarized ourselves with its customs, language, culture, and history. Part I prepares you to begin your journey into food chemistry.

Chapter 1—Why Do We Eat? provides a bit of the history of why eating is necessary for life. We will learn that food is the fuel that our bodies need to live. We will also learn the language of food science: the chemical vocabulary used to describe the molecular structure of each type of food. We can then understand how foods differ in the way they are used for energy. Food is also necessary for growth, and our nutrients must provide us with the raw materials for growth. We will learn that carbohydrates, fats, and proteins, three important nutrient groups, have distinct chemical compositions that govern how they are assembled and broken down. In later chapters, we will use this information to see that cooking food retraces many of these transformations.

In **Chapter 2—Making Sense of Our Senses**, we will learn how our bodies utilize sensory cues to form the integrated sensory response of flavor. We will examine the various sensory signals that food provides and explore how smell and taste work in unison to create a food's flavor, while texture, sight, and sound play important but secondary roles. Here we will also learn about molecular and cellular communication. How is it that sweet taste receptors recognize sugar molecules and that aromas from baking bread bind to receptors in our noses? Signals from taste and smell receptors are sent to the brain, where they are integrated and assigned a flavor.

Equipped with the chemical vocabulary to describe our nutrients and the rules for chemical communication presented in part I, we are ready to move on to Part II, which shows how each cooking technique can transform raw foods into wonderful dishes that provide us with both energy and building blocks for growth.

CHAPTER

Why Do We Eat? 1

In this chapter, we begin our exploration of the underlying chemical, biochemical, and physical principles that enable us to understand food and its transformation in the kitchen and in our bodies. Our journey starts in **Section 1: Food for Survival**, with a discussion of the body's need for fuel and raw materials for growth and answers the fundamental question of why eating is necessary for life. Food is the fuel we burn, and calories are how scientists measure the total energy released when a food undergoes complete combustion. In **Section 2: The Three Basic Food Groups**, we develop the chemical vocabulary to describe the unique molecular structures of the three primary macronutrient food groups: carbohydrates, fats, and proteins. Cooking and digestion can transform raw food's complex molecular structures into simpler forms we call building blocks. Carbohydrates are broken down into simple sugars, fats are broken down into fatty acids, and proteins are broken down into amino acids. In these forms, they can be metabolized by the body, and the energy captured can be used later to provide energy measured in calories. When necessary, the building blocks can be used to make new materials needed by a growing body. In **Section 3: Molecular Control of Metabolism and Appetite**, we introduce the key signaling molecules that regulate our food intake at the cellular and organism level.

Appetite is a means by which we adjust the amount and type of food we consume. The chemical signals, known as hormones, send information throughout our bodies at each step.

SECTION 1: FOOD FOR SURVIVAL

Like breathing and drinking water, eating is one of the basic means of sustaining human life. Survival guides refer to a "Rule of Threes," in which humans can (on average) survive three minutes without oxygen, three days without water, and three weeks without food.[1] But *why* do we need to eat? At a basic level, our bodies require energy to attend to the business of living. Each contraction of our heart, each electrical signal sent or received by the brain, and each exchange of gases in our lungs require energy. The fats, carbohydrates, and proteins in our food undergo a controlled combustion reaction in which they are oxidized like the wood in a campfire, ultimately combining with molecular oxygen in the air to produce carbon dioxide and water. In the process, energy is released. The light and heat produced by combustion in a campfire is not terribly useful in our bodies, so instead our bodies store the energy from the controlled combustion of food in the form of chemical energy. This chemical energy is stored in the bonds of molecules such as adenosine triphosphate (ATP), a molecule often referred to as the energy currency of the cell. Figure 1.1 shows how the stored chemical energy is released when the bond holding the terminal phosphate in ATP is broken, forming the more stable products, adenosine diphosphate, ADP, and phosphate. To replenish the stores of ATP, the body uses the energy released by the controlled combustion of food to make more ATP. Energy is also stored in other molecules such as fat deposits, glycogen, or ultimately, and only as a last resort, the proteins that make up our bodies. As the grim survival statistics reveal, these energy reserves usually sustain life for three weeks without food.

When food is consumed, it is broken down step by step during digestion, and the energy that is released is captured throughout the process. Many of these steps are controlled by protein molecules called enzymes. Each type of enzyme binds very specifically to one type of molecule known as a substrate and transforms it into a particular product. The enzyme catalyzes the transformation of substrate to product, which means the reaction occurs faster than it would otherwise and can occur at body temperature. "Catabolism" is the term given to the collection of processes necessary for the complete breakdown of food into its final products, carbon dioxide and water. To give you an idea of the complexity of the catabolic processes essential for digestion, at least eighteen different enzymes are needed to take the simplest food source, glucose, to carbon dioxide and water. Ten of these reactions occur in one compartment of our cells known as the cytoplasm, and the rest occur in a different part of our cells, the mitochondria. Carbohydrates, proteins, and fat each follow their own catabolic breakdown

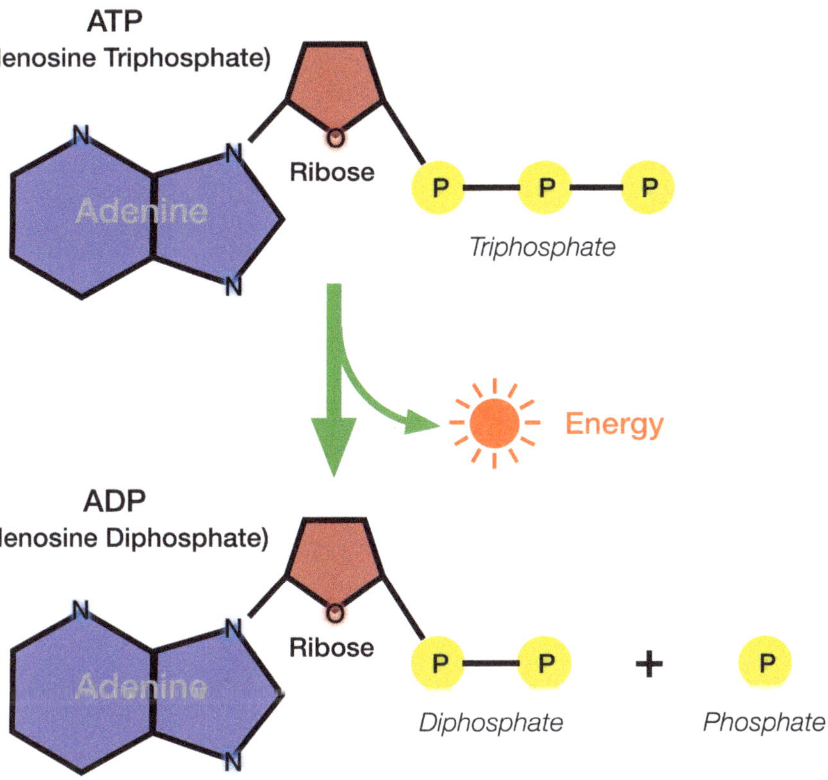

(a)

Figure 1.1 Energy Is Stored in the Bonds of Adenosine Triphosphate (ATP)

Energy is released when ATP is broken down into ADP and phosphate. The key to understanding this process lies in the complex molecular structure of ATP.

a. Cartoons help us to understand the structure. The three component parts of ATP are the adenine in purple, shown as a hexagon fused to a pentagon, the pentagonal ribose sugar in red, and the triphosphate group, shown as three yellow circles. In converting ATP to ADP, the adenine and ribose do not change, but the bond holding the terminal phosphate group is broken. In this process, energy is released and used by the body.

Figure 1.1 continues on the following page.

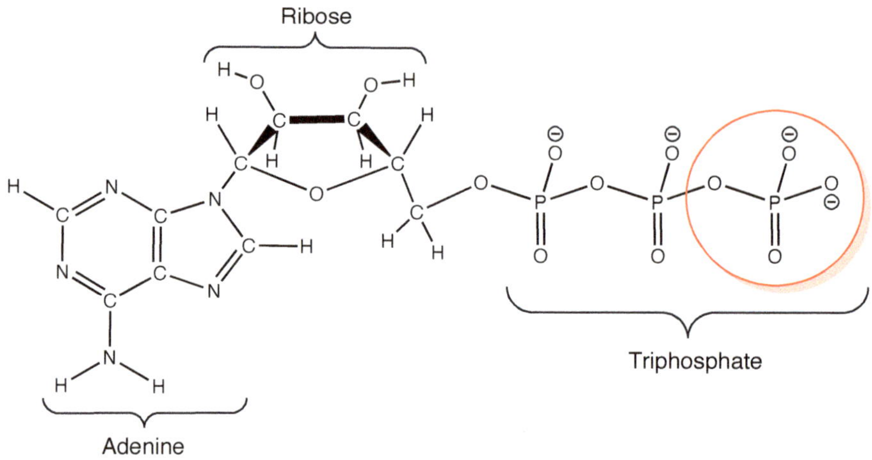

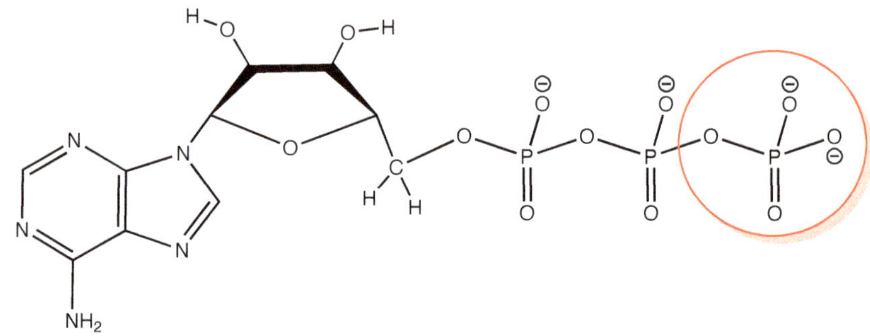

(b)

Figure 1.1 *(Continued)*

b. Inspection of the more complete molecular structures is the next step in our understanding of why ATP transformation is accompanied by energy release. We can see in the top image an explicit molecular structure of ATP in which all the atoms that make up this molecule are drawn. The atoms are connected to each other through bonds that are either single (one line) or double (two lines). The three phosphate groups are each made up of a single phosphorous atom (P) surrounded by four oxygens (O), some of which have a negative charge. ATP has a total of four negative charges quite close to one another and since negative charges repel each other, this arrangement is quite repulsive. The phosphodiester bond holding the red-circled phosphate group to the rest of the molecule will be broken when ADP is made, and the reduction of concentrated charge from -4 to -3 is one reason why so much energy is released as the product molecules of ADP and P are more stable. The bottom image is an implicit molecular structure of ATP that omits the drawing of some carbons and hydrogens and that chemists use as a shorthand.

pathways. As stated previously, an essential goal of catabolism is to produce energy, which is stored as chemical energy in the bonds of the ATP molecule.

In 1990, for the first time, food manufacturers in the United States were required to include product-specific labeling with "per serving" caloric and nutritional information on all prepackaged food.[2] Today we also see this information on menus and recipes, but how is it determined, and is it accurate? One

BOX 1.1 Note on Chemical Structure Drawings

We will use a shorthand that chemists have developed to represent the structure of complex molecules known as the implicit chemical structure. For simplicity, implicit chemical structures do not draw all the carbon or hydrogen atoms. A carbon should be placed at the intersection of two lines if no other atom is explicitly drawn. Since all carbon atoms should make four bonds (we say carbon has a valence of 4), hydrogens will fill in if a particular carbon in the structure does not have four bonds. So, for example, if you see a carbon with only two bonds drawn explicitly, you should know that there will be two additional bonds to two hydrogen atoms. Another feature of a molecule important for its biological function is its shape. Most food molecules have a three-dimensional shape that is important to the way they bind to other molecules like receptor proteins or enzymes. To draw a three-dimensional shape on the two-dimensional page, we use darker lines or wedges to indicate that this part of the molecule is pointed toward you and dashed lines or wedges to indicate that this part of the molecule points away from you.

way scientists measure calories is to burn, or combust, a sample of material using a bomb calorimeter. Here a sample is ignited in a closed container containing oxygen. The heat released causes the temperature to rise in a surrounding water bath, which can be precisely measured. Since one calorie is defined as the amount of energy needed to heat one kilogram of water 1°C, the temperature rise can be related to the number of calories. This method is time-consuming and expensive, so food manufacturers use a much more straightforward but less precise process known as the Atwater method. More than a century ago, William Atwater, a chemistry professor at Wesleyan University, determined that, on average, each gram of protein or carbohydrate produced four calories, and each gram of fat produced nine calories.[3] He reasoned that if one multiplies the number of grams of proteins plus carbohydrates by four and the number of grams of fats by nine, the sum would provide an estimate of the calorie content of the food. Today the Atwater method is used by food manufacturers, restaurant chains, and most food establishments to calculate calories, even though it is an estimate and can be misleading. For example, some foods, such as nuts, might contain calories that are not absorbed.[4] Individuals can also differ significantly in how their metabolic systems are regulated, which may cause them to burn fuels differently. Exactly how much energy a specific food will provide to a particular individual will depend on hundreds of individual metabolic and environmental factors that are nearly impossible to predict. Our bodies extract energy from food in much more complicated ways than the precise bomb calorimeters or estimated Atwater methods used to determine the calorie content.

Table 1–1 **Typical Cellular Turnover in the Human Body**

Examples of Cell Type	Turnover (average, varies)
Cells of stomach, intestines	5 days
Taste buds on tongue	2 weeks
Epidermal cells of skin	1 month
Hair	3 years (men)
	6 years (women)
Bone	10 years
Lens of eye	Do not replicate
Neurons in cerebral cortex	Do not replicate

Did you know that the calorie reported on a food label in the United States is not a calorie but a kilocalorie (kcal)? The prefix *kilo-* means 1,000 of whatever unit is being measured. So my granola bar, (mis)labeled as providing 200 calories, will produce not 200 calories but 200,000 calories, or 200 kcal, of energy when combusted in the laboratory. This is an old convention for package labeling that is unfortunately still used in the United States, but it is scientifically incorrect.

A balanced diet typically means a distribution of foods that includes macronutrients—proteins, fats, and carbohydrates, and smaller amounts of micronutrients—vitamins, minerals, and fiber. This distribution is necessary for homeostasis, defined as a state of balance among all the body systems needed for the body to survive and function correctly.[5] Package labels provide us with the breakdown in terms of a percentage of how much of the minimum daily requirement (MDR) is met for each type of food. This is important because food also provides us with the raw materials that provide building block molecules to make new cells or replace aging ones through a collection of processes known as anabolism. Once again, these processes occur in multiple steps, each controlled by an enzyme, some of which are the same enzymes working in reverse. You need the right raw materials for homeostasis to give you the right building blocks. Growing muscles need amino acids that you can get from eating proteins. However, your body is brilliant at using alternative strategies if you don't provide it with the most preferred type of building blocks. In addition to being burned for energy, the dietary fats you eat can be broken down into basic units that are recombined to create new cell membranes or synthesize hormones. Carbohydrates or sugars that you consume will be broken down into basic units that can recombine into the signaling molecules that serve as hormones or coat our cells and direct how one cell interacts with another.

Our anabolic needs might best be illustrated by the data in Table 1–1, the typical turnover or lifetime for different types of cells in the human body. Replacement of dead cells requires raw materials to build up the more complex molecules—anabolism. The replacement cycle for some cells is on the order of days (stomach) or weeks (taste buds). Other cells last for years (hair and bone). Some cells are *never* replaced, such as the cells in the lens of your eye or neurons in your cerebral cortex. The cells in the lens of your eyes were there when you were born and will be there when you die, so treat them with care! Replication occurs at different rates at different stages of your growth, at different times of the year, and can depend on your sex, age, environment, and what you had for dinner last night.

Calories and nutritional balance are two parameters we often track and use to plan meals and make dietary choices. The breakdown of food for energy (catabolism) is constantly being balanced by the need to grow or replace cells (anabolism). Being healthy and being in homeostasis depends on the proper balance between those two processes.

How are all these processes that will often go in opposite directions managed? Many regulators read the body's needs and direct traffic as necessary. Children need to consume raw materials that will enable the growth of healthy bones and teeth. After puberty, we need to consume raw materials that will provide for the possibility of reproduction. When we are running a marathon, the primary need is for energy, so stores of energy in the liver and muscle and fat are mobilized. One type of regulator is the growth hormone estrogen. Many of these shifts in how our bodies require and use food are directed by chemical signals known as hormones.

SECTION 2: WHAT ARE THE THREE BASIC FOOD GROUPS?

The three essential macronutrient food groups, carbohydrates, fats, and proteins, are needed in large quantities to provide the cellular energy for us to live (catabolism) and the raw materials for growth (anabolism.) In this section, we will learn the chemical nature of each of these macronutrients. We will focus on the ways they can be broken down into building blocks that have unique chemical structures and built up into more complex structures held together by unique chemical bonds that link one building block to the next.

All about Carbohydrates

Who does not crave a slice of freshly baked bread? Starting our discussion of metabolism with carbohydrates makes sense because it is one of the most fundamental ways we (and most living organisms) get energy. A carbohydrate is a molecule made up of atoms of carbon (C), hydrogen (H), and oxygen (O). When we write the molecule's chemical formula, subscripts tell us how many of each atom is present. Glucose is the most important simple sugar and an excellent place to start. One glucose molecule comprises 6 carbon atoms, 12 hydrogen atoms, and 6 oxygen atoms and is described using the chemical formula $C_6H_{12}O_6$. Further inspection shows us that the ratio of atoms in glucose is 1 C for every 2 H and every 1 O, or CH_2O. Here we see how the word *carbo*[C] *hydrate*[H_2O] makes sense. Water has the chemical formula H_2O; when water is attached to something else, the resulting molecule is described as a hydrate. This ratio of atoms, CH_2O, is called the empirical formula.

The monosaccharides glucose, fructose, and galactose all have the same chemical formula, $C_6H_{12}O_6$. Figure 1.2 shows that, despite this, these molecules all have different structures—that is, the atoms are connected to one another in different ways. In this figure, we are using the implicit structural representation as described in Box 1.1, where C atoms, though not drawn, exist at the intersection of two lines, and H atoms, again not always drawn, are there to make sure every C has a valence of four. Confirm for yourself that each monosaccharide has the same chemical formula, $C_6H_{12}O_6$. The different connectivities lead to different three-dimensional shapes of the molecules, as shown in Figure 1.3. Polarity is another important property of molecules that derives from whether the valence (outermost) electrons of the molecule are balanced. In this book, we use cloud overlays to represent the polarities or surface charge densities. Figure 1.3 also shows that the different connectivities of glucose and fructose have resulted in different surface charge densities. Receptors or transport proteins recognize molecules by their shape and polarity features and can tell the difference.

When two simple sugars bind together, a water molecule is lost, and the resultant molecule is called a disaccharide. The new bond joining the two simple sugars, the glycosidic bond, can be formed in either alpha (α) or beta (ß) geometry, as shown in Figure 1.2. The disaccharide lactose from milk is made from galactose and glucose with a ß linkage. The disaccharide sucrose from ordinary table sugar is a combination of glucose and fructose with an α linkage.

More complex carbohydrates are called polysaccharides, and Figure 1.2 shows two common examples. Cellulose, the most abundant biopolymer on earth, consists of glucose units bound together using a ß geometry for the glycosidic bond. These linkages produce long linear chains and allow cellulose to act as a structural polymer to build cell walls in green plants. Humans lack the enzyme that would enable them to break down the ß glycosidic bond in cellulose, so we cannot digest grass or cotton. Starch is another polysaccharide made from repeating units of glucose, but here the linkages between the sugars use the α geometry, which results in a very different bent shape.

Digestion of disaccharides and starch in our bodies begins with the breakdown of the glycosidic bond. Enzymes known as salivary amylases start that process as food enters your mouth. Naturally occurring sugars such as lactose from milk, maltose from grains, and fructose from fruit have been part of the human diet forever. Today highly processed sugars such as those from cane, beet, and corn are widely available, inexpensive, and a growing percentage of our daily caloric intake.

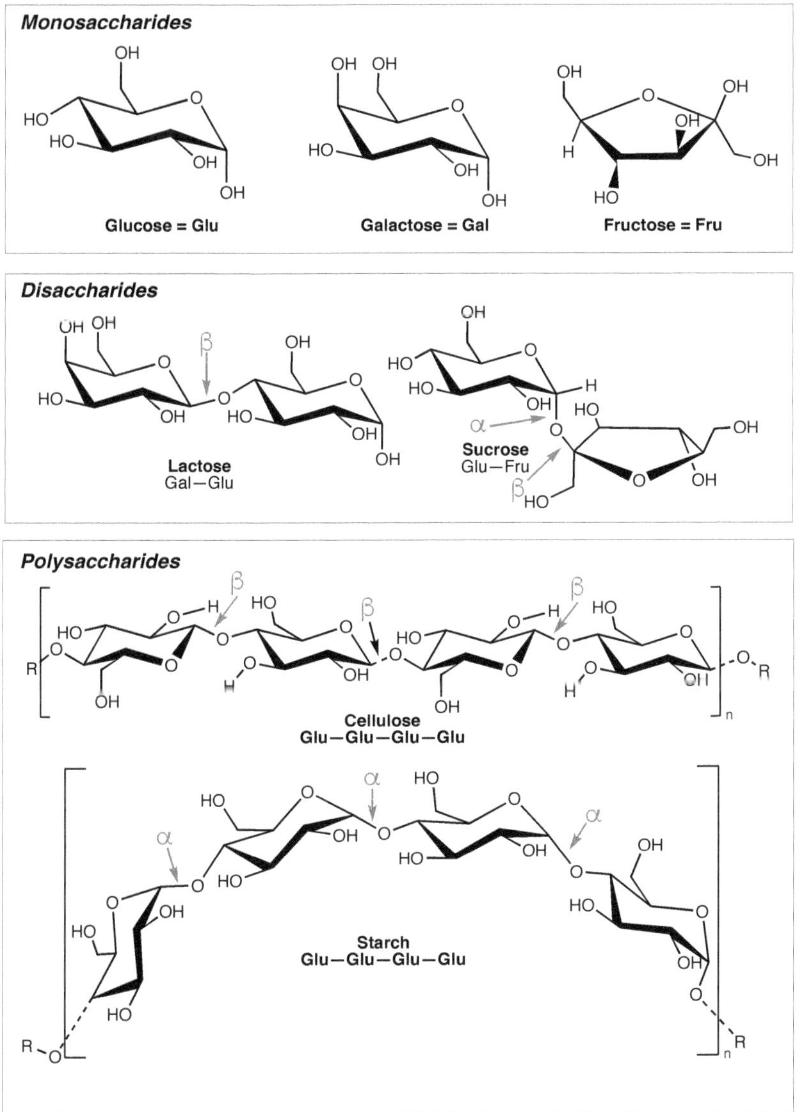

Figure 1.2 Simple Sugars Are the Building Blocks of Carbohydrates

The monosaccharides glucose (Glu), galactose (Gal), and fructose (Fru) all have the same chemical formula $C_6H_{12}O_6$ but different chemical structures. The disaccharides lactose and sucrose are formed by joining two monosaccharides, Gal-Glu for lactose and Glu-Fru for sucrose. The bond holding the simple sugars together is called a glycosidic bond. The geometry of the linkage bond is important to the shape and fate of the more complex sugars. The two geometries are distinguished by the Greek letter α (two sugars connected bottom to top) or β (two sugars connected side to side.) Notice how the linkage geometries result in a straight molecular geometry for cellulose (β linkage) and a curved molecular geometry for starch (α linkage).

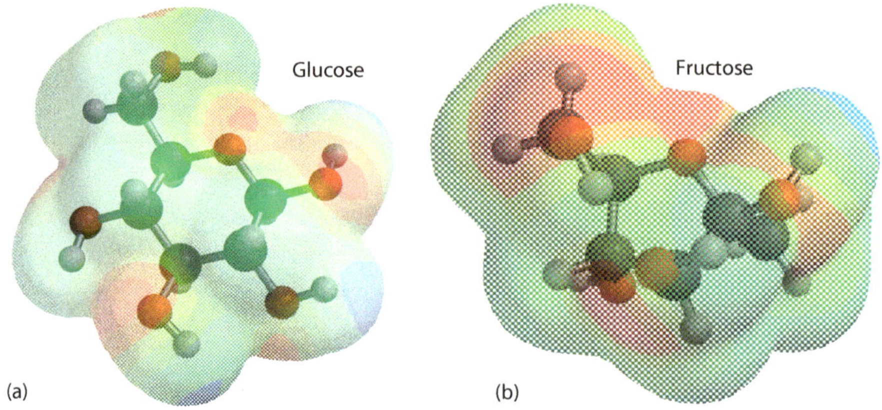

Figure 1.3 Glucose and Fructose Have Different Geometries and Surface Charge Distributions

Despite their identical chemical formulas, $C_6H_{12}O_6$, the monosaccharides glucose and fructose have different shapes and charge distributions. The ball and stick models for the molecules show atoms as color-coded balls (C black, H white, and O red), and the bonds represented as sticks. We can see that the atoms are connected differently in the two molecules: glucose has a six-membered hexagon ring, and fructose has a five-membered pentagon ring. Different polarities of the two molecules can be seen by the colored cloud overlay, in which the surface charge has been calculated. Areas with slightly greater electron density are colored red (partially negative), areas with slightly less electron density are colored blue (positive,) and areas with balanced electron density are shaded green (neutral.) Though the net charge on the molecule is zero, the overlay shows that the electrons are not distributed evenly for these molecules.

All about Fats

Fat is a word with many meanings in popular usage (having too much flabby tissue, corpulent, obese, or even stupid such as in "fathead"). To a chemist, fat is a class of macronutrient molecules known as triacylglycerols (TAGs). The rather complex structure of TAG can be taken apart and shown to have a short three carbon glycerol backbone to which three fatty acid molecules are attached.

Before considering the molecular structure of fats in detail, we should first acknowledge a most important property from a culinary perspective: they do not mix with water. Anyone who has ever made a vinaigrette knows oil and vinegar (which is mostly water) when mixed do not spontaneously stay mixed. This is because vinegar (water) is a polar molecule, and oil is a nonpolar molecule, but what do these terms mean at the molecular level? Figure 1.4 uses surface charge images to show how the electrons are distributed in the molecule. In water, we see unevenly distributed electrons as represented by the red (electron-rich) and blue (electron-poor) surfaces. By contrast, oils, such as triacylglycerol, have evenly distributed electrons, represented by the green surface. The same chemical principles responsible for the lack of solubility of oil and water also result in fats and, more generally, lipids being insoluble in water. In general, polar molecules interact well with other polar molecules, while

BOX 1.2 Carbohydrates: Ten Fun Facts

1. For sugar and carbohydrates, the ratio of atoms is ($C*H_2O$).
2. Most sugars that are digested start out as a 6-carbon sugar, or hexose. It is also possible to digest five carbon sugars, or pentoses. The suffix *-ose* in chemistry means "sugar." The prefixes *hex-* in hexose and *pent-* in pentose tell you the number of carbons (6 and 5, respectively).
3. A synonym for sugar is saccharide, from Latin *saccharum*, "sugar." Simple *mono*saccharides have one six carbon sugar unit (glucose, fructose). More complex *di*saccharides have two six carbon sugar units (sucrose, lactose.) Large *poly*saccharides have many sugar units (glycogen, starch, cellulose).
4. In confectionary kitchens, chefs will play with different sugars that have different melting temperatures and different properties, sometimes even adding enzymes to a solid sugar that will break it down into simpler sugars with a low melting point (think of the liquid sugar centers of chocolate-covered cherries).
5. In most organisms, including humans, a pathway known as glycolysis (pronounced *gly-col-y-sis*) breaks down a six carbon sugar to two carbon sugars in ten steps—none of which require oxygen; that is, they are anaerobic. In aerobic organisms like us, a second pathway known as the citric acid cycle then breaks down the three carbon sugar even further to CO_2 and H_2O in eight steps that require oxygen.
6. Glycolysis is an ancient anaerobic pathway that evolved in living organisms three billion years ago when the atmosphere had very little oxygen. These early cells developed pathways to use sugars for energy, even when there was not enough oxygen in the atmosphere to "burn them."
7. Whether or not an organism can digest a complex polysaccharide depends on the nature of the linkage between two simple sugars. Each type of linkage is broken down by its own individual enzyme, and humans have the enzyme to digest the linkage in glycogen or starch but do not have the enzyme to digest the linkage in cellulose. Meanwhile, the stomachs of both termites and cows contain microbes that do have the enzyme, thus enabling termites to eat wood and cows to eat grass.
8. Our gut microbiome helps us by facilitating absorption of sugars from the gut by breaking down complex carbs to simpler carbs.
9. Sugar processing disorders—for example, diabetes—can sometimes be managed by controlling the amount and type of carbohydrate consumed. Glycemic indices (GIs) tell us how quickly the carb is digested. Foods with high GIs are digested quickly, producing unhealthy spikes in blood sugars. Low GI foods like oatmeal are digested slowly with the help of our gut microbiome. These keep the blood sugar at a more constant lower level and gets rid of unhealthy spikes.
10. **Though variations exist, when burned all the way to CO_2 and water, carbohydrates produce an average of 6.3 ATP/carbon atom.**

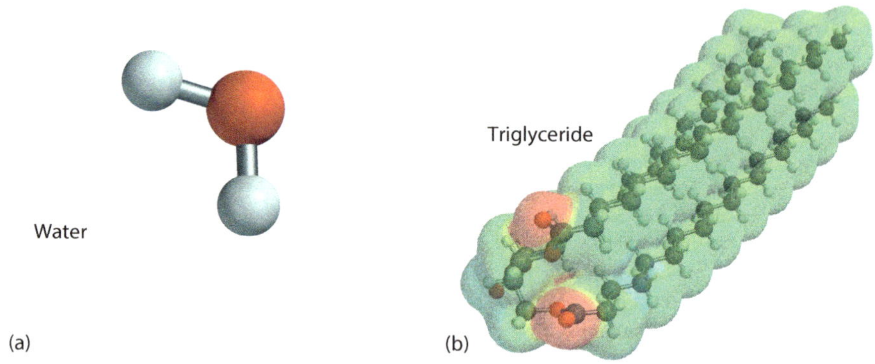

Figure 1.4 Calculated Surface Charge Distributions for Polar and Nonpolar Molecules

The surface charge image for a water molecule (H_2O) shows that the electron density is not balanced. There is extra electron density around the oxygen atom (red shading) and less electron density around the hydrogen atoms (blue shading.) Compare this to the surface charge image for a triglyceride molecule, which shows that the electron density is evenly distributed (green shading.) These key differences make water polar and the triglycerides that make oil nonpolar. This is why water and oil don't mix.

nonpolar molecules interact well with other nonpolar molecules. This is expressed in a phrase you may hear in the lab or kitchen: "Like likes like."

Fatty acids, as shown by the explicit molecular structures (all atoms) and ball and stick drawings in Figure 1.5, are made from long carbon-hydrogen chains connected to a carboxylic acid (-COOH). These chains are typically longer than four carbon atoms (C4) and range up to about twenty carbon atoms (C20), but they always contain an even number of carbon atoms. The carbon atoms are connected to each other by mostly single bonds, the simplest and weakest type of C-C bond. If *all* the C-C bonds in a fatty acid are single bonds, the molecule is described as "saturated," as is the case for palmitic acid (C16 fatty acid found in palm trees) and stearic acid (C18 found in cows).

It is also possible for some of the C-C bonds in a fatty acid to be stronger double bonds, as shown for the oleic acid molecule. The term "unsaturated" describes a fatty acid with double bonds. Oleic acid, the predominant fatty acid in olive oil, has eighteen carbons, just like stearic acid. However, it has one double bond and two fewer hydrogen atoms, so it is called "monounsaturated." The double bond between atoms 9 and 10 is relatively rigid, producing a kink in the molecule's structure. Molecules with more than one double bond are called "polyunsaturated." If this all sounds familiar to you, it is because human overconsumption of one or the other type of fat can lead to disease. Polyunsaturated and monounsaturated fats are often called healthy fats. Overconsumption of saturated fats (butter, red meat) can lead to cardiovascular disease and strokes.

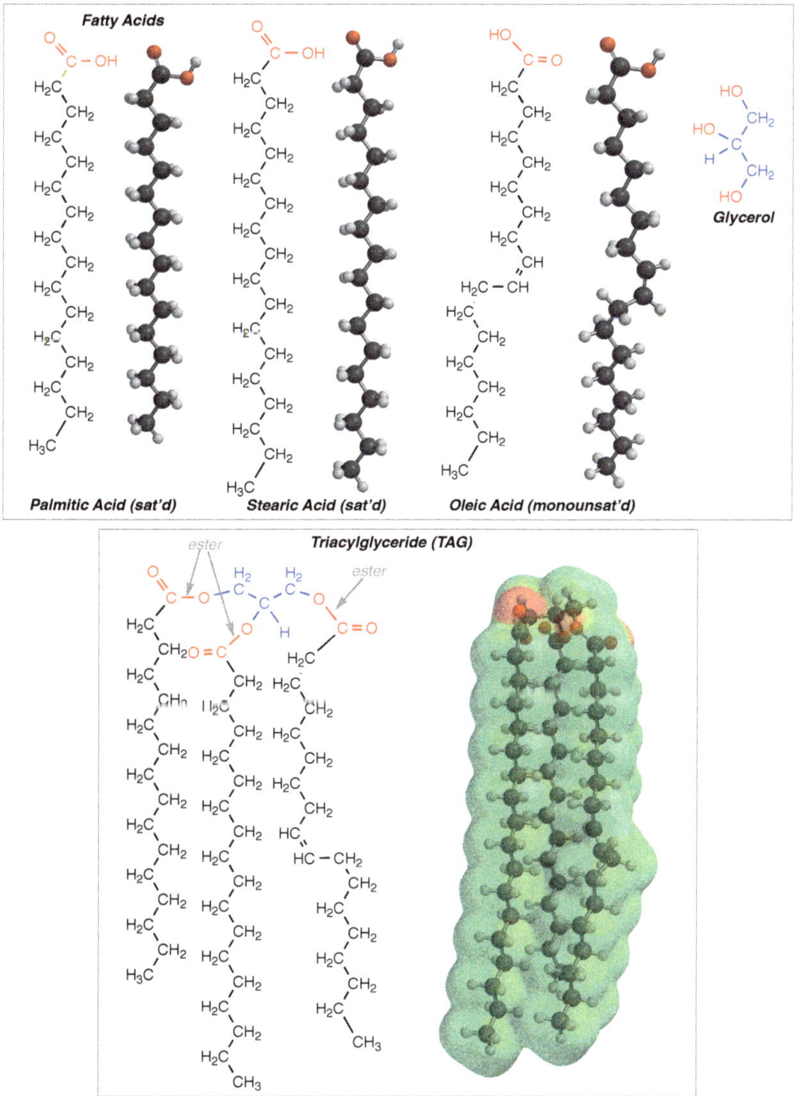

Figure 1.5 Fatty Acids and Fats

Fatty acids are molecules with a long chain of carbon atoms connected to a carboxylic acid group (-COOH.) When there are no carbon-carbon double bonds, the fatty acid is called saturated, such as the sixteen-carbon palmitic acid (C16) and the eighteen-carbon stearic acid (C18). A fatty acid with one or more double bonds is described as unsaturated, as shown by the oleic acid, an eighteen-carbon fatty acid with one double bond (C18 monounsaturated). Fatty acids with more than one carbon-carbon double bond are characterized as polyunsaturated. Three fatty acids can bind to one glycerol molecule to make a triacylglyceride (TAG) or fat through an ester linkage. The polar surface rendering of the TAG shows the long nonpolar (green) tail of the molecule and the polar head (red and blue) of the glycerol.

Fats and oils are made up of triacylglycerol (TAG) molecules. The TAG has a backbone glycerol molecule to which three fatty acids are connected. Glycerol, shown in the top right side of Figure 1.5, is a simple molecule with three carbon atoms each connected to an alcohol group (-OH). To make TAG, the oxygen of each alcohol group in glycerol binds to the carbon of the carboxylic acid of each fatty acid and a water molecule is released. The newly made carbon-oxygen bond is known as an ester bond. Adding a water molecule back again can break down this bond, a process known as hydrolysis. In our bodies, enzymes known as lipases help to begin the digestion of fats by hydrolysis.

Figure 1.5 also shows the surface charge for TAG; the predominance of the neutral green surfaces predicts the poor solubility of lipids in water. This might have presented a problem for the utilization of fats by the body. However, our bodies have adapted clever ways of moving these insoluble but necessary nutrients around to where they are needed. For the insoluble TAG from our diet or from fat tissues (adipose tissue) to enter the bloodstream, it must be bound to carrier proteins such as low-density lipoprotein (LDL) and high-density lipoprotein (HDL). Typically, the fat is transported first to our liver, the command center for fat utilization, and then to cells to be burned for energy or to cells that need it as a raw material for growth.

Once the TAG enters the cell, the fatty acid groups are first removed from the glycerol backbone by hydrolysis of the ester bond. Then the free fatty acid (remember, fatty acids always contain an even number of carbons) is broken down in multiple steps, with each stage creating a two carbon unit (known as an acetyl group). As in all combustion, the final products are CO_2 and H_2O and energy in the form of ATP. What happens to the CO_2 produced? Usually, CO_2 travels through the blood to the lungs, where it is exhaled into the atmosphere.

All about Proteins

Of all the food groups you can eat, proteins are the worst sources of energy at only 5 kcal/carbon. However, they are very important because when broken down, they can provide amino acids, which are the building blocks for new proteins necessary for growth. Protein-rich foods like meat, fish, eggs, nuts, and dairy have the added benefit of making us feel full for longer as it takes longer to be broken down, absorbed, and transported to sites of protein synthesis.

Just as simple sugars are the building blocks of carbohydrates and glycerol and fatty acids are the building blocks of fats, proteins and peptides (small proteins) use amino acids as their building blocks. As the name suggests, every amino acid has an amino group ($-NH_2$) on one end and a carboxylic acid group (-COOH) on the other. A central carbon atom known as the alpha carbon ($C\alpha$) is between the amino's N and the acid's C. The $C\alpha$ is also connected to a side group expressed in shorthand as -R and in all but one amino acid, a hydrogen atom. The side group can be any of the twenty unique atoms or molecules. The

BOX 1.3 Fats: Ten Fun Facts

1. Fatty acids are added to a three carbon glycerol to make a triacylglycerol (TAG) that acts like a storage packet for the fatty acid.
2. Fats are not soluble in water, so we say fat and water are "immiscible." You know that oil floats on water. This insolubility presents a big challenge for digestion. Having a fat layer floating on top of the primarily aqueous environment of the body would mean that enzymes or receptors would have to travel across the fat layer and out again to work.
3. The body's way of dealing with the insolubility of fats is to package them in little or big suitcases (HDL, LDL, and chylomicrons) when they need to be shuttled around in the body, say, from the intestinal mucosa where they are absorbed to the liver, which is command central for deciding what to do with the fat. Should it be processed into cholesterol or used as a raw material for steroid hormones or cell membranes, or should it be burned by the liver and other cells for energy, or should it be stored in adipose tissue?
4. During digestion, lipase enzymes must first break down a C-O bond between the glycerol and the fatty acid. This hydrolysis begins in the intestines as bile acids and lipases work to disassemble the TAG.
5. The blood of some individuals has high levels of fat or their derivatives, which leads to cardiovascular problems that lead to strokes and heart attacks. Blood levels of cholesterol, both the good HDL and the bad LDL, along with the free TAG, are used to monitor the lipid levels of the blood.
6. The melting temperature of fat is related to both the number of double bonds (the degree of unsaturation) and the length of the long carbon chains. Fats with more double bonds (higher unsaturation), such as vegetable, corn, and nut oils, behave like liquids. Fats with fewer double bonds (higher saturation), such as butter, lard, and tallow, behave like solids.
7. Historically, the fats humans consumed were linked to where they lived. Plant oils were widely consumed in areas with abundant palm and olive trees. In other areas, humans turned to domesticated animals and made butter and ghee from cows.
8. Human enzymes cannot make double bonds beyond a certain distance from the carboxylic acid group, even though some of these unsaturated acids are necessary for good health. Fortunately, other species can do this, so for good health humans must consume fats from those animals, such as oil from the liver of codfish, to get the fatty acids with those characteristics.
9. The age of olive oil, an unprocessed fruit oil, can be determined by the positions of the fatty acids on the glycerol backbone. When just harvested, all the fat exists as TAG (triacylglycerol), with all three fatty acids connected to the glycerol. As the oil ages, it slowly degrades. First, the end fatty acid bond is hydrolyzed to make 1,2 DAG (diacylglycerol). Then the fatty acid at position 2 migrates to position 3 to make 1,3 DAG. Oil with a high 1,3 DAG is stale and should be thrown away or used to make soap.
10. **Though variations exist, when burned to CO_2 and water, fats produce an average of 8.1 ATP/carbon atom, the most of any food source.**

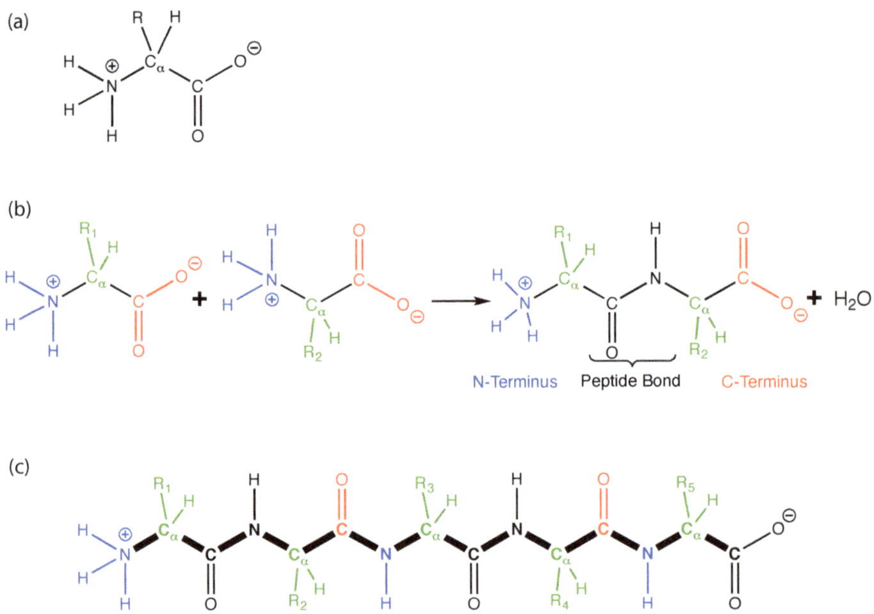

Figure 1.6 Amino Acids Are the Building Blocks of Proteins

Amino acids contain a three-atom backbone of a nitrogen atom, a central carbon atom called the α-carbon, and a carbon from the carboxylic acid group, as shown in Figure 1.6a. The α-carbon also bonds to a hydrogen atom and a side group called an R group. The R group can be any of the atoms or molecules that make up the twenty standard amino acids. At neutral pH, amino acids have a positive charge on the amino end and a negative charge on the carboxy end. In Figure 1.6b, a peptide bond is formed when two amino acids come together to create an amide bond and a water molecule is lost. The dipeptide formed will have a positively charged N-terminus and a negatively charged C-terminus. Figure 1.6c shows a pentapeptide made from five amino acids joined together by four peptide bonds (notice the five R groups labeled R_1-R_5). By convention, the N-terminus is usually shown on the left and the C-terminus on the right. The heavy black line traces the atoms of the peptide's backbone. When the number of amino acids exceeds approximately fifty, we call this a protein.

side group of the amino acid proline is unique in that it bends back and makes a second bond to the Cα, which is why there is no hydrogen atom on this Cα. The basic structure of an amino acid building block is shown in Figure 1.6**a**. Each amino acid has a unique chemical behavior due to the variation in the -R group size, shape, polarity, and reactivity. Each is distinguished by unique three-letter and one-letter codes, as shown in Figure 1.7.

Figure 1.6b shows the bond known as a peptide bond that is formed when we link two amino acids together. In making that bond, a water molecule is lost—a process known as dehydration. The synthesis of proteins—polymers of amino acids—occurs on a cell's ribosome by a wonderfully complex and highly regulated process. However, as we consider the transformation of proteins that happens during cooking, our focus will be on breaking down the weaker bonds that hold the three-dimensional structures of the proteins together rather than breaking the peptide bonds that give us energy and new raw materials.

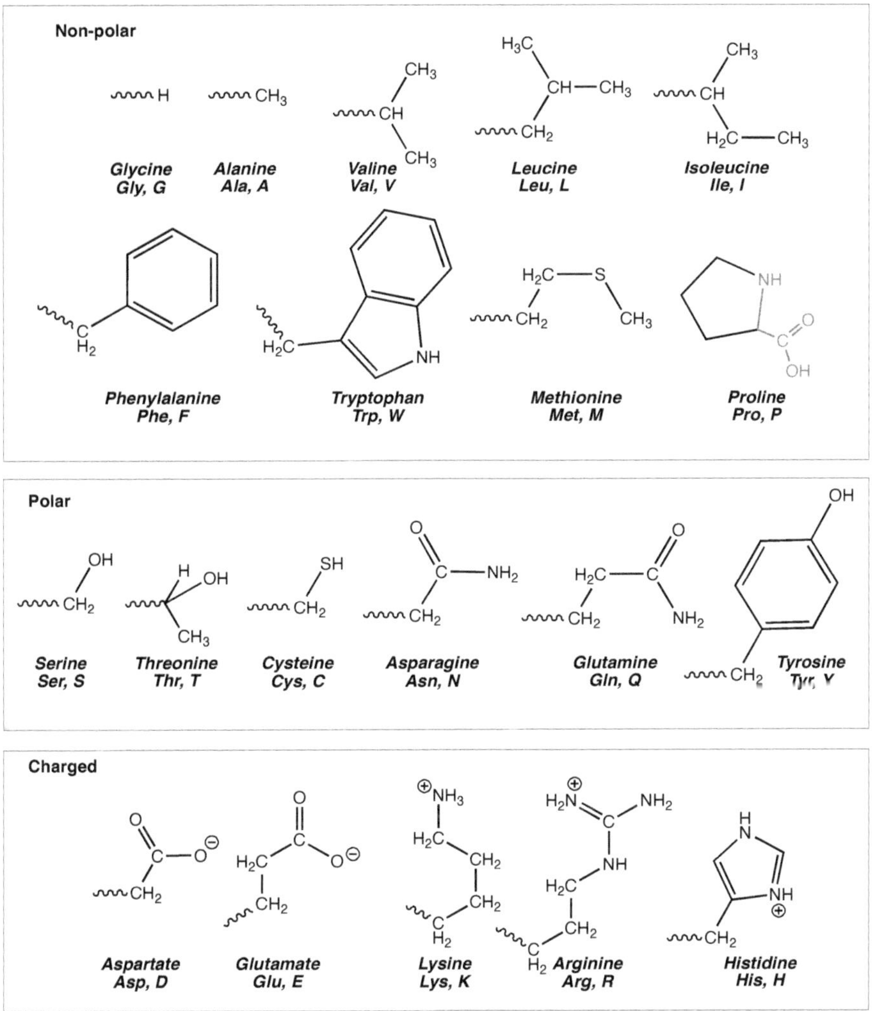

Figure 1.7 Names, Side Group Structures, and Classifications for the Twenty Amino Acids

Each amino acid has a unique full name, a three-letter code, and a one-letter code based on the identity of the side group. Skeletal structures for the side groups, which can be as simple as an H atom for glycine, are shown above the name. The squiggly line represents the bond that connects the side group through the α-carbon to the rest of the amino acid. For simplicity, these common atoms are not shown. The amino acids can be organized into three groups based on the solvent preferences of the side group: nonpolar, where the side group prefers nonpolar solvents; polar, where the side groups have N, O, or S atoms that can participate in hydrogen bonding to water; and charged, where the side groups can have positively or negatively charged atoms that prefer to be solvated in water. These charges depend on pH; the likely charges at pH 7 are shown here.

Proteins are described in many ways, but usually we start by considering the amino acid sequence, beginning with the amino terminal (H_2N-) or N terminal and moving toward the carboxy terminal (-COOH) or C terminal. It is customary to use the one-letter codes for the amino acids to describe the individual amino acids. The 20 amino acids can be thought of as 20 letters from which hundreds of thousands of protein "words" can be made. Thus, a small peptide made up of five different amino acids histidine (H), glutamate (E), alanine (A), arginine (R), and threonine (T)—would be written H_2N-HEART-COOH. Often, for simplification, we leave off the H_2N- and -COOH, since, by convention, the amino terminal is always written on the left and the carboxyl on the right, or in this case, we would report the pentapeptide as "HEART." Just as in spelling, the sequence, or order, of amino acids matters. The same five amino acids we used to spell the pentapeptide HEART (H, E, A, R, and T) could have been used to spell EARTH or HARTE—or any of the 120 combinations of those five letters. When making a word, each combination has a different meaning. In the same way, different amino acid sequences will result in different shapes and functions.

A protein's shape, function, and stability depend on the composition (how many of each amino acid), the sequence of amino acids, and the total number of amino acids, which can be anywhere from about 50 to 500,000. In this chapter, we will soon learn about four signaling hormones made from amino acids: insulin (53 amino acids), glucagon (29 amino acids), ghrelin (28 amino acids), and leptin (146 amino acids). Humans have many different peptides or protein hormones, each with a different sequence and structure, and each is used to send a different signal and effect a different physiological response. These hormones are recognized by receptor proteins on the surface of cells to which they bind to transmit their signal. The sequences and structures for the four signaling hormones noted in this chapter are shown in Figure 1.8.

However, acting as signaling molecules is not the only thing proteins do in the body. They are necessary structural materials for making muscle, connective tissue, hair, and skin. Another special class of proteins is enzymes, as mentioned earlier in the chapter, and these proteins have the unique job of acting as catalysts to make reactions go faster than they would otherwise. Without enzymes, a candy bar might take a hundred years to be broken down into parts. Who can wait a hundred years to digest a candy bar? With the enzymes in your saliva, the breakdown of that candy bar begins in your mouth. It continues with other enzymes in your stomach, intestines, and, eventually, in your cells—taking minutes or hours but certainly not years.

SECTION 3: MOLECULAR CONTROL OF METABOLISM AND APPETITE

The phrase, "Do you eat to live or live to eat?," has been attributed to such luminaries as the fifth-century BC philosopher Socrates, the eighteenth-century

BOX 1.4 Proteins: Ten Fun Facts

1. Proteins are polymers of amino acids connected by a bond known as a peptide bond.
2. Twenty different amino acids are used to make proteins. They all have an amino group, a middle carbon, and a carboxylic acid group but differ in their substituent -R groups.
3. Nine of these amino acids, called essential amino acids, cannot be made by our bodies but must be consumed in our diet. They are histidine, isoleucine, leucine, lysine, methionine, phenylalanine, threonine, tryptophan, and valine. All can be provided by protein-rich diets that include meat, fish, eggs, beans, lentils, dried peas, tofu, tempeh, nuts, and seeds.
4. Phenylketonuria (PKU) is a rare inherited disorder that causes the body's amino acid phenylalanine to build up. PKU is caused by a defect in the gene that helps create the enzyme needed to break down phenylalanine. Newborns are screened at birth for PKU. If the disorder is present, the babies must have a low-protein diet excluding phenylalanine.
5. Proteins are unfolded, or "denatured," during cooking (the egg goes from clear to solid). You can denature proteins with heat, salt, or acid. All these techniques are used in various food preparation techniques.
6. Digestive enzymes rapidly break down proteins by hydrolysis or by adding a water molecule to the peptide bond. Since a digestive enzyme itself IS a protein, how does it manage not to break itself apart into its component pieces? Some enzymes avoid this problem by being made in an inactive precursor molecule that is only activated when food is consumed.
7. Protein powders are food additives that can be added to milkshakes. Because food additives are not regulated as thoroughly as other types of consumables, these proteins can be from various sources, such as soybeans, eggs, milk, potatoes, or even hemp. Nevertheless, these distinctions will not be found on a food label, which may present a problem for people with food allergies.
8. Protein buildup and breakdown must be properly balanced and adjusted to an individual's needs. Many human diseases occur because of a breakdown in this regulation, with too much protein being created or too little being degraded.
9. The word *protein* is derived from the Greek word *proteios*, meaning "primary" or "first," which is certainly understandable given the importance of proteins for nutrition.
10. **Though variations exist, when burned to CO_2 and water, proteins produce an average of 5 ATP/carbon atom, the least of any food source.**

Figure 1.8 Four Hormones That Regulate Metabolism and Appetite

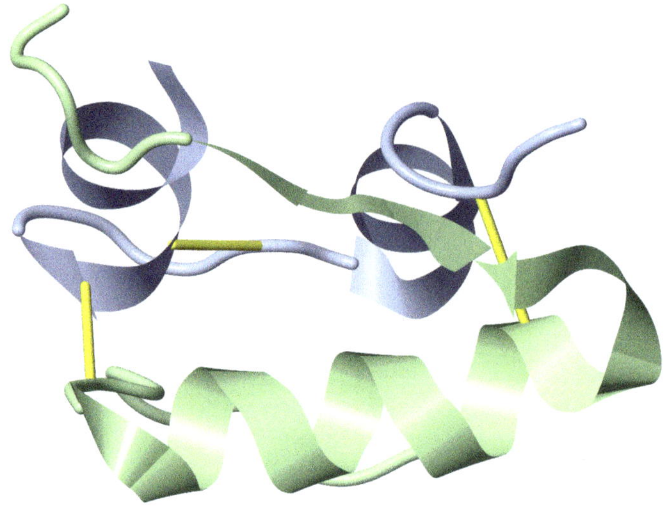

Figure 1.8a shows the important peptide hormone insulin, with 788 atoms that are distributed between the 23 amino acids of the A chain.

GIVEQCCTSICSLYQLENAYCNA

and the 30 amino acids of the B chain.

FVNQHLCGSHLVEALYLVCGERGFFYTPKT

The cartoon image shows the backbone tracing for the A chain as a blue ribbon containing two helices joined by a disordered region. The cartoon image for the B chain is green with a more extended helix and a disordered region at the C terminus. The two chains are held together by two disulfide bonds (S-S) made from the sulfur-containing side groups of the amino acid cysteine, C, and shown by yellow rods.

Image was created in FirstGlance using coordinates for pdbID: 1TRZ for insulin.

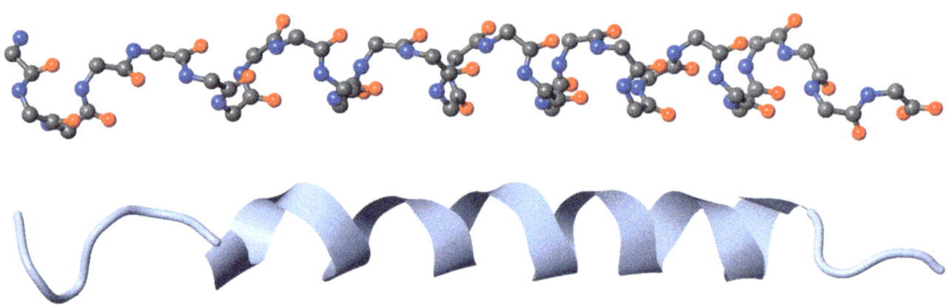

Figure 1.8b shows glucagon, the peptide hormone that signals when blood glucose levels are low, which has 471 atoms from 29 amino acids with the sequence.

HSQGTFTSDYSKYLDSRRAQDFVQWLMNT

Two different representations are shown. The first image shows nonhydrogen and backbone atoms displayed using a ball-and-stick model. For simplicity, no side group atoms are shown. A coil or helix is suggested as the backbone atoms are traced from left to right. The second image is a cartoon representation of the backbone, where no atoms are shown at all, but a ribbon traces the backbone, here a single helix. Since these ribbon diagrams make it so much easier to see the structure of a polypeptide, it is the style most biochemists use when describing proteins.

Images were created in FirstGlance using coordinates for pdbID: 1GCN for glucagon.

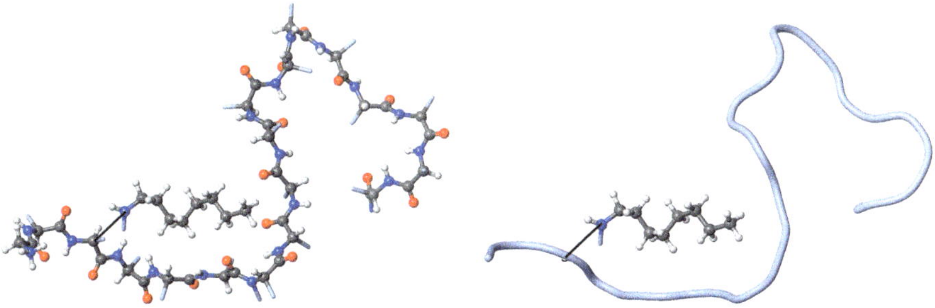

Figure 1.8c shows ghrelin, the hunger hormone, which is the smallest hormone we have discussed, with only 28 amino acids. Using the single-letter amino acid designation, we can represent the entire sequence as:

GSS*FLSPEHQRVQQRKESKKPPALKQPR

While this is a small hormone, it still has 320 atoms. A picture with all atoms represented would be difficult to interpret. A simplified structure is shown on the left, which draws just the backbone atoms and leaves off the side groups. This allows us to see that in addition to the amino acids, there is an 8-carbon fatty acid, octanoic acid, attached to the side group of the third amino acid, serine, S (marked with an asterisk in the sequence.) Further simplification by replacing the backbone atoms with a solid thread or ribbon allows us to focus on the structure without being distracted by the atoms. The final image is referred to as a cartoon representation. We can see that the ghrelin structure is disordered in that it has no regular helices or strands, but it forms a pocket for the fatty acid group.

Images were created in FirstGlance using coordinates for pdbID: 6H3E for ghrelin.

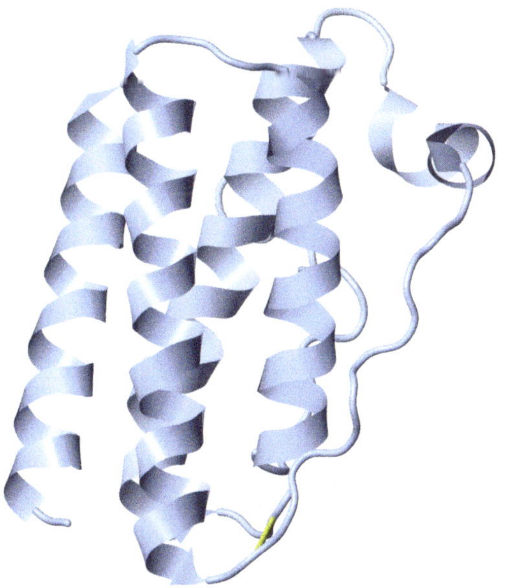

Figure 1.8d shows the appetite-suppressing hormone leptin, with 130 amino acids, is the largest of the four hormones referred to in this chapter. We will not attempt to show a structure with the positions of all the 1074 atoms. This ribbon diagram shows the backbone with five helices and some disordered regions or loops connecting the helices. The helices are stabilized by the hydrogen bonds described in 1.7b. The packing of one helix next to another helix is stabilized by weak noncovalent interactions.

Image was created in FirstGlance using coordinates for pdbID: 1AX8.

inventor and statesman Benjamin Franklin, and the twentieth-century fitness pioneer Jack LaLanne. We can consider whether to focus on the metabolic level (eating to live) or the gastronomic level (living to eat). An even more satisfactory response is that we do not need to choose and that both considerations are important in healthy eating. Fortunately, systems exist to regulate both processes. We will first consider how our bodies regulate food uptake in our cells (eating to live) and then consider how our bodies regulate our appetite: what we choose to eat, how much, and when (living to eat.) Both processes and many others are controlled by signaling molecules known as hormones. Hormones are molecules made by one cell or organ in the body and secreted into the blood, where they travel to remote cells or organs where they turn on, speed up, slow down, or stop some biological process. Hormones can be as small as the nitric oxide molecule NO, medium-sized like the steroid hormone estrogen, or larger, such as the protein hormones described here that regulate sugar metabolism and our appetite.

Regulating Metabolism at the Cellular Level

All the macronutrient groups can be metabolized to generate energy, but carbohydrates, mainly glucose, are our bodies' most fundamental energy source. Blood sugar uptake and utilization depend on two hormones: insulin and glucagon. Working together, they **keep blood sugar levels within safe ranges**. Insulin prevents blood sugar levels from rising too high (hyperglycemia), while glucagon stops blood sugar levels from dropping too low (hypoglycemia).

Insulin signals that sugar is on its way.

The most important hormone that regulates sugar metabolism is the protein hormone insulin. Insulin is a small protein made in your pancreas and secreted into the bloodstream to signal cells throughout your body to get ready to digest sugar. Insulin will bind to receptors on the surface of cells and turn on the cellular processes needed to transport glucose into the cell. Just the smell of food causes insulin secretion from your pancreas. Humans whose pancreas cannot produce insulin have Type 1 diabetes, which usually manifests early in life. These individuals must regularly inject insulin to help them digest sugars. Type 2 diabetes has several different origins, but the most common type results from the inability of cells to respond to insulin signals. This disease manifests later in life and can often be managed by choosing a diet that avoids processed foods high in carbohydrates. It is also essential for people with diabetes to monitor their blood sugar and, if necessary, take medication if their blood sugar gets too high or too low.

Insulin will influence how or if our cells take up sugar from the blood, but this depends on the type of sugar and the type of cell. We have seen that glucose and fructose have the same molecular formula ($C_6H_{12}O_6$) but different structures. Receptors and enzymes to which they will bind will recognize these differences. When insulin is present, receptors throughout the body transport glucose from the blood into the cell. By contrast, due to its unique fructose transporter, the liver is the only organ that can absorb fructose from the blood, and the transport is not under insulin control. In the first step of digestion, most cells use the enzyme hexokinase, which binds to all six carbon sugars (hexoses) and starts the process of digestion by adding a phosphate group to the sugar. Hexokinase exhibits feedback control, which means that if the product of the reaction builds up, the enzyme can be slowed down or stopped. By contrast, once fructose enters the liver, it encounters a unique liver enzyme, fructokinase, that is fast-acting and has no feedback control to shut it down if product levels are too high.[6] There is much debate in the medical community regarding the relationship between the consumption of high fructose corn syrup (HFCS) and obesity and other lifestyle diseases.[7,8] The different fates and pathways for these two simple sugars and the various ways they are taken up by cells and processed highlight the enormous complexity and fine-tuning managed by hormones such as insulin.

A cartoon representation of insulin is shown in Figure 1.8a. The cartoon hides the complexity that would be necessary to visualize the 788 atoms. The ribbon diagram traces the backbone N-Cα-C atoms of the hormone and allows us to see that insulin is made up of two peptide chains: a blue ribbon for chain A and a green ribbon for chain B. The chains are held together by yellow rods, representing sulfur-sulfur bonds from atoms in the side groups of two cysteine -R groups. The ribbons loop into a three-dimensional structure known as a helix that is stabilized by many weak bonds between side groups and stronger hydrogen bonds within the molecule. It is these weaker bonds that can be easily broken by heat in cooking.

Glucagon signals that our blood supply of sugar is low.

When blood sugar drops, the transport processes shut down, and metabolism slows. But what if you suddenly need the energy to, say, race after a toddler about to step onto a busy street? Our bodies have several systems for providing energy quickly in an emergency. The most well known is adrenaline, a steroid hormone secreted by the adrenal gland. Another less well-known system for giving your body a quick energy source is the peptide hormone glucagon (glucose is gone!). Like insulin, glucagon is a small protein made in and secreted by the pancreas. Cells that respond to the glucagon signal activate enzymes needed to break down a storage form of sugar known as glycogen in liver and muscle cells. Individuals with glycogen storage diseases must take injections of this hormone.

The protein structure for glucagon is shown in Figure 1.8b. Here we have two images: a ball and stick image that traces only the backbone atoms and the carbonyl O and a bottom image that shows the ribbon tracing of the backbone and that glucagon is just a single helix. The ball and stick image allows us to see that the red oxygen atoms point along the helix axis, which allows them to make hydrogen bonds with the hydrogens that attach to the blue N atoms. These are the bonds that are most important in making this helical structure stable.

Insulin to glucagon ratios might influence obesity.

The ratio of the hormones insulin and glucagon is an important parameter in a model of obesity known as the Carbohydrate-Insulin Model (CIM). This model challenges the dominant Energy Balance Model (EBM), which considers obesity an energy balance disorder. EBM states that when people take in more calories than they burn, they become fat. Sedentary lifestyles exacerbate the energy imbalance. EBM does not explain that despite an explosion of low-calorie foods and gym memberships, global obesity has tripled since 1976.[9]

A CIM of obesity postulates that it is not just the calorie content of our food but the type of calories, with a particular focus on carbohydrates and high glycemic index foods.[10] Hormones and many other factors control the partition of the substrates we eat.[11] The two peptide hormones described here, insulin and glucagon, play a most important role in directing the traffic of metabolism for homeostasis.

Regulating Appetite at the Level of the Organism

Hunger and satiety are signals that, together, govern our appetite. Living to eat rather than eating to live might be one way of distinguishing our metabolic needs from the excitement and pleasure we wish to create by cooking dishes that taste good and are satisfying. Once again, we find that hormones are the messengers that carry the physiological signals between our digestive system, our adipose tissues, and our brains.

Ghrelin signals that you need to eat.

The peptide hormone ghrelin is made and stored in cells in the digestive system. Like a lookout, it constantly monitors the status of the stomach. An empty stomach causes the activation of ghrelin by attaching a small eight carbon long fatty acid (octanoic acid) to the hormone, after which it is secreted into the bloodstream, where it is quick to act at two different sites, the stomach and the brain. One set of receptors in your stomach binds to ghrelin and initiates borborygmi, that embarrassing gurgling and rumbling of an empty stomach that is essentially your stomach announcing to the world, “THIS GUY IS HUNGRY!” Ghrelin also travels through the bloodstream to the brain, where it binds to a receptor in the hunger control center of your brain, the hypothalamus. The signal initiated by ghrelin is the sense of being hungry—your body telling you, "I am HUNGRY! I must EAT!"

Two different images of the hormone ghrelin are shown in Figure 1.8c. In the first image, though the protein side groups are not shown explicitly, the other atoms still look rather complex, and it is hard to see what is going on. Figure 1.8c also shows a simplified image in which the backbone atoms have been replaced with a thread and now, it can be seen that there is a fatty acid group attached to the peptide. It also shows that the peptide does not have a regular structure. We call this structure disordered. It might be the case that once ghrelin binds to a receptor or to the membrane, it does acquire a nice three-dimensional structure.

Leptin signals that you are full.

Once you have eaten enough to satisfy your metabolic needs, the slower-acting hormone leptin is secreted from adipose tissue (fat storage cells). It travels through the blood and up to the brain, where it binds to other receptors in the hypothalamus to signal, "SATISFIED! STOP EATING!" Besides sending signals to your brain, leptin increases the metabolic rate of other cells to ensure your cells will burn the food eaten. This signaling hormone, shown, is a critical enzyme in weight control. Studies have found that the leptin receptors in many people have become insensitive to leptin—a condition associated with obesity. No matter how much is eaten, the brain never receives the SATISFIED signal. Not surprisingly, leptin is a potential therapeutic target for those individuals who are chronically obese.[12] Figure 1.8d shows a ribbon diagram of leptin, with four helices packed into what is known as a four-helix bundle, a structure found in many proteins.

SECTION 4: HEALTHY EATING BEYOND CARBS, FATS, AND PROTEINS

While the chosen focus of this book is the transformations of the macronutrients (carbohydrates, fats, and proteins) during cooking, it would be negligent to ignore the other important consumables necessary to keep you healthy. We need to drink water and consume fiber regularly in our healthy diet. Micronutrients, vitamins, and minerals are other molecules we must consume in small quantities to keep us healthy. The Daily Value (DV) and Recommended Daily Allowance (RDA) for many macronutrients[13] and micronutrients are listed in Table 1–2. A goal is to choose a varied and healthy diet that contains balanced macro- and micronutrients. However, this is often difficult, so people sometimes supplement their diets with multivitamins.

As hard as it is to believe, more than half of our body is made up of water, and replenishing its supply is a critical need, as dehydration can cause death in a few short days. Water excreted by our kidneys and bowels helps us eliminate the toxins produced as by-products of metabolism. In addition to helping us eliminate toxins, sweat is necessary for keeping our body temperature regular. Water protects sensitive tissues like our brain and helps lubricate our joints. Adults should drink about 64 ounces of water daily to stay healthy and hydrated. Precise recommendations are hard to pin down, as your water needs vary significantly by activity level, gender, and general health. Of course, if you are in a hot environment or exercising, you must drink more.

Keeping "regular" is a little-appreciated part of our general health until we are not. In addition to water, we turn to fiber or roughage. Roughage includes those parts of the plants and grains that we eat that our body does not digest or absorb. These undigested components of our food have several benefits for our body, including regularizing our bowel movements and improving gut health. Insoluble fibers are found in whole wheat flour and bran, seeds and nuts, cauliflower, green beans, and potatoes. In its soluble form, obtained from grains such as oats and barley, or legumes such as peas and beans, or fruits such as apples and some citrus, or vegetables such as carrots, fiber also helps lower cholesterol and slow the uptake of sugar into our cells, both of which are good things. How much do we need? Though many factors influence an individual's requirement for fiber—the daily value of 28 grams is used for food labeling.

Micronutrients are needed in minimal amounts but are critically necessary to prevent disease and develop strong bones and muscles. Since our body contains tissues and fluids mainly composed of fat or water, it is essential to have some water-soluble micronutrients and some fat-soluble ones. The most important vitamins typically include the following six:

- Fat-soluble vitamin A is found in orange and red fruits and vegetables and is essential for vision and immune health.
- B vitamins are found in leafy greens and fortified complex carbohydrates. They are considered most important for mental health, general energy levels, and growth, which is why they are crucial ingredients in prenatal vitamins. Eight B vitamins, including folate, thiamine, niacin, and biotin, are essential, and most Americans do not get enough from their diet.
- Water-soluble vitamin C is found in citrus and vegetables and helps promote healing by preventing tissue damage from oxidative free radicals.
- Fat-soluble vitamin D is essential for bone health and regulation of calcium and potassium—best obtained by a daily dose of sunshine.
- Fat-soluble vitamin E, found in many oils and fruits such as avocados, is another molecule that fights oxidative damage and protects against cell damage and muscle and skin health.
- Vitamin K, also in leafy greens, is necessary for blood clotting.

Table 1–2 **Minimum Requirements for Macro- and Micronutrients and Fiber**

FOOD NUTRIENT	DAILY RECOMMENDED VALUE	NOTES
Macronutrients		**FDA Changes 2016**
Protein	50 g	None
Carbohydrate (total)	275 g	Down from 300 g
Added sugar	50 g	New category
Fat	78 g	Up from 65 g
Saturated fat	20 g	None
Cholesterol	300 mg	None
Dietary Fiber	28 g	Up from 25 g
Micronutrients		
Vitamin A	700 μg (women) 900 μg (men)	Fat soluble
Vitamin B	1.1 mg	All water soluble
B1, B2, B6		Very little variation by gender
B3	14 mg	
B5	5 mg	
Biotin	30 μg	
Folic Acid	400 μg	
B12	2.4 μg	
Vitamin C	65 mg (women) 95 mg (men)	Water soluble
Vitamin D	15 μg	Fat soluble Older adults take 20 μg
Vitamin E	15 mg	Fat soluble
Vitamin K	120 μg (men) 90 μg (women)	
Calcium	1,000–1,200 mg	Water soluble Higher end for older adults
Iron	8 mg 0–27 mg	Varies by age and gender 0 mg for infants up to 6 months 27 mg for pregnant women
Zinc	11 mg (men) 8 mg (women)	
Sodium	1,500 mg	Typical diet: 3,300 mg
Potassium	2,600–3,400 mg	Varies with gender and age Typical diet: 2,300 mg

Minerals are another class of micronutrients that are typically ionized metals such as calcium (Ca^{+2}), iron (Fe^{+2}), or zinc (Zn^{+2}). These three minerals are often included in multivitamin supplements as our diets sometimes fall short of the RDA. Their jobs and sources are listed below.

- Calcium is found in dairy foods and spinach, soy, and rhubarb and is essential for healthy bone growth, blood pressure regulation, and hormone secretion.
- Iron is an integral component in the protein hemoglobin. It helps get oxygen into our cells by binding the oxygen in the lungs and carrying it through the blood to cells where it is needed for the complete combustion of our foods. In its absence, we feel lethargic and can develop anemia. Iron is found in red meat and leafy greens.
- Zinc is found in red meats, poultry, and nuts and helps our memory and our bodies fight colds by boosting our immune system.

Other common minerals not found in supplements are sodium (Na^{+1}), which is found in salt and essential in the control of blood pressure and keeping the proper fluid balances outside of cells and which we typically consume way too much of, and potassium (K^{+1}), which is found in dried apricots, papayas, and avocados and is the counterpart of sodium in the control of fluids inside of cells. Since the contraction of muscles depends on these fluid balances, potassium is essential for muscle health. Sodium and potassium have significant and opposite effects on our bodies, and many diseases, especially cardiovascular disease, are linked to getting the balance of these out of whack. While high salt (sodium) levels can lead to high blood pressure, which can, in turn, lead to a heart attack, high potassium relaxes blood vessels and helps us excrete salt. In a well-balanced diet, we eat more potassium than sodium, but the reverse is true for most Americans.

SUMMARY OF CHAPTER 1

Food provides our fundamental needs for energy and raw materials for growth. Metabolism is the series of steps that break down complex molecules in our food to provide energy stored as chemical energy in molecules like ATP. The energy content of food is measured in calories. Anabolism is the process by which simpler building blocks create the more complex molecules we need for structure and growth. In this chapter, the complex chemical structures of our three macronutrient groups (carbohydrates, fats, and proteins) were shown in each case to be made from simple molecular building blocks (sugars, fatty acids, and amino acids). These simple building blocks combine to create more complex molecules using unique chemical bonds—a glycosidic bond for sugars, an ester bond for connecting fatty acids to glycerol, and a peptide bond for proteins. Coordinating food uptake and storage at the cellular level and hunger and satiety at the organism level is a complex job regulated by signaling molecules known as hormones. These hormones have unique structures and show exquisite selectivity to the molecules they bind.

Along with the host of micronutrients, a balanced diet is necessary for homeostasis.

In the next chapter, we will explore the chemistry and biochemistry behind our senses and learn how sensory inputs guide our food selection and appetite for food. Knowledge of the chemical structure of food presented here will provide insight into the cooking processes in which these complex structures are broken down in Part II—Transformation. But first, let's make sense of our senses.

CHAPTER 2

Making Sense of Our Senses

We often define what we eat by the unique molecules that make up our foods, such as carbohydrates, fats, and proteins. We carbo-load before a race, smother our mashed potatoes with Dad's gravy made from luscious fat drippings, and crave the protein in a hamburger off the grill. Those carbohydrate, fat, and protein molecules interact with all five of our sensory organs. **Section 1: The Fivefold Architecture of Flavor**, describes how our brains integrate the signals into an overall sensory picture. For many substances related to flavor, these interactions rely on a fundamental principle in chemistry known as molecular recognition: the ability of one molecule to recognize and bind to another molecule based on its size, charge, polarity, or other molecular properties.

Section 2: Our Five Senses, explains how this principle applies to sensory science. Molecular recognition describes how a taste bud or olfactory receptor can select one molecule from the hundred different types of molecules in food and bind to it tightly. The sensory molecule, usually called a receptor, is part of a sensory cell that connects with a nerve that sends a signal to the brain. All aroma receptors and some taste receptors work on this principle of molecular recognition. Other taste cells work by responding to a difference in the concentration of a molecule or ion between the inside and outside of the cell. Concentration differences destabilize the cell, and channels open to let the molecule or ion in a greater concentration outside the cell flow in. For sight and sound, our eyes and ears respond to visible light or sound that arises from the food, either when we look at it or when we listen to the sounds made when chewing. You will learn that touch plays a much more significant role than you can imagine in flavor appreciation. Each sense has its circuitry that connects the sensory organ to the brain, and complex interactions can exist between them.

In **Section 3: Food Pairing**, we see that when we decide what foods complement each other, we use principles that go back to the knowledge of molecular compositions. We like food pairings with standard aroma components, such as fruity or floral aromas. We also enjoy contrasting tastes like sweet and sour and different textures like smooth and crunchy.

SECTION 1: THE FIVEFOLD ARCHITECTURE OF FLAVOR

Close your eyes, and think of the most delicious thing you have ever tasted. What can you remember of the aroma or the texture? Was it sweet or savory? Did it last on your tongue or seem, like cotton candy, to evaporate in your mouth? Was it appealing visually, or did it challenge your expectations? For most of us, taste, like breathing, is something we often take for granted and rarely think carefully about while eating. Tasting involves the neural response to the foods we eat, from both the taste buds in our mouths and throats to particular sweet, sour, bitter, salty, and umami molecules *and* the trigeminal sensors on our tongues that detect temperature/spiciness/texture. When combined with our olfactory response to the aroma of our food, taste results in a perception of food that we describe as the flavor of a particular food. Our purpose in this chapter is to develop our appreciation of food by examining, one by one, the different sensory contributions that our five senses contribute to our appreciation of and ability to manipulate the taste and flavor of the foods we eat and the dishes we prepare.

Figure 2.1 highlights the different regions of the human brain that are the primary locations for touch, taste, vision, hearing, and smell. The part of the brain that responds to smell, the olfactory bulb, is buried underneath the cerebrum, toward the front of the brain. Taste signals travel first to the medulla and then to the thalamus. The sensory signals are then processed using regions of the brain that respond to memory, learning, and novelty.[1,2] It is interesting to consider how much brain "real estate" is dedicated to each of the five senses and compare it to the sensory organ itself. The nose is certainly not the largest sensory organ. Still, the olfactory bulb takes up a large portion of the brain, suggesting its incredible importance to overall conscious or unconscious perception. Similarly, our eyes, ears, and tongues are relatively modest-sized

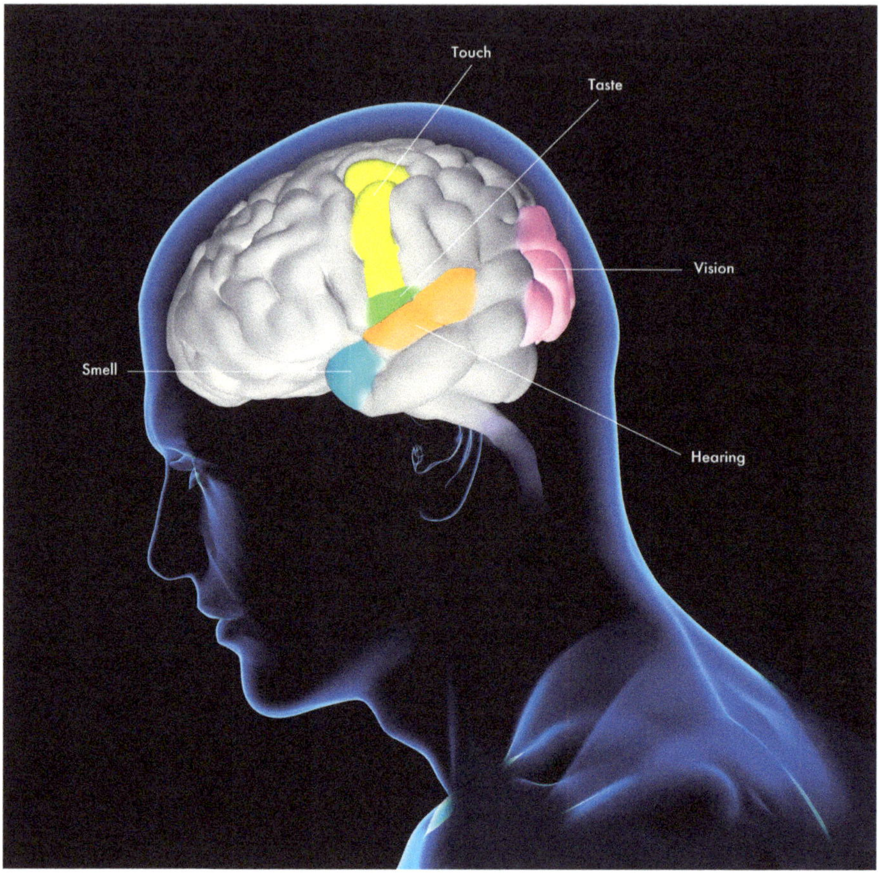

Figure 2.1 Sensory Map of the Brain

The five senses are mapped onto the physical areas of the brain primarily responsible for that perception. Our five senses, taste, sight, touch, hearing, and smell, all contribute to what we think about as the flavor of our foods.

organs but occupy quite a bit of brain real estate that connects to other brain regions, leading to complex integrations.

Our perception of flavor depends on combined information from all five senses. For example, the aroma of a roasted pecan pie stimulates our odor-sensing cells directly through the nose (orthonasal smelling) and through odor molecules that find their way to the nose from the back of the oral cavity (retronasal smelling.) In addition, the golden-brown color, sweet taste, gooey and chunky texture, and crunching sound we hear as we bite into the pecans are all integrated within the brain to give us the delicious flavor of a pecan pie.[3]

SECTION 2: OUR FIVE SENSES

The Senses of Smell and Taste

Taste and smell are the two senses most directly associated with flavor. We all know that smell or olfaction occurs in the nose, while taste or gustation occurs in the mouth. Both result from detecting a particular chemical in the food by a receptor, and that receptor causes a nerve signal to pass to the brain. That is where the similarity ends. As shown in Figure 2.2, chemicals that enter the nose, called odorants, must be gaseous or volatile particles that can travel through the air. Aromas, or smell, account for about 80% of the flavor of food; smells can cause us to either salivate, which prepares our digestive system before the food ever enters our mouth, or feel repelled, as a disgusting odor makes it nearly impossible to consume a particular food. When it is necessary to consume food or medicine that we find repulsive, what do we do? We hold our noses. In the mouth, non-volatile molecules called tastants wash through the mouth while we are chewing and swallowing. The tastants signal receptor cells that exist in clusters within taste buds or in surveillance cells in the trigeminal area of the tongue. These cells "read" the food and alert the brain: salty, hot, or spicy. Together with the receptors in the nose, the receptors in the mouth help us to determine the answer to the question: Will this food kill me? Is it safe to eat? After that, we can freely enjoy the food's hedonic (pleasure) aspects.

"Olfaction" is the scientific term for the sense of smell.

While humans have about 1,000 genes for olfactory receptors, only about 300 become receptor proteins in the nose; the remaining genes remain dormant in our DNA and are never expressed. We do not know why only about 30% of our olfactory repertoire is used, but even with this diminished capacity, our sense of smell is the primary factor in taste. Dogs, on the other hand (or paw), not only have forty times more receptor capacity in their noses than we do, but they use more of the genes they possess. Even with our more limited olfactory repertoire, humans can recognize more than 10,000 different odorants. How is that possible? The scientists Linda Buck and Richard Axel studied the olfactory response.[4] They learned that while the olfactory receptors bind to one and only one type of odorant, the odorant molecules themselves can bind to many different types of olfactory receptors. Each odorant effectively produces a "bar code" signal, a map of which receptors responded and how strong the response was, which the brain eventually interprets. To illustrate this, consider heptanoic acid, a volatile odorant molecule responsible for the smell of rancid oil. Figure 2.3 shows that heptanoic acid produces a mild response from receptor A, a strong response from receptors C, D, and F, and no response from the

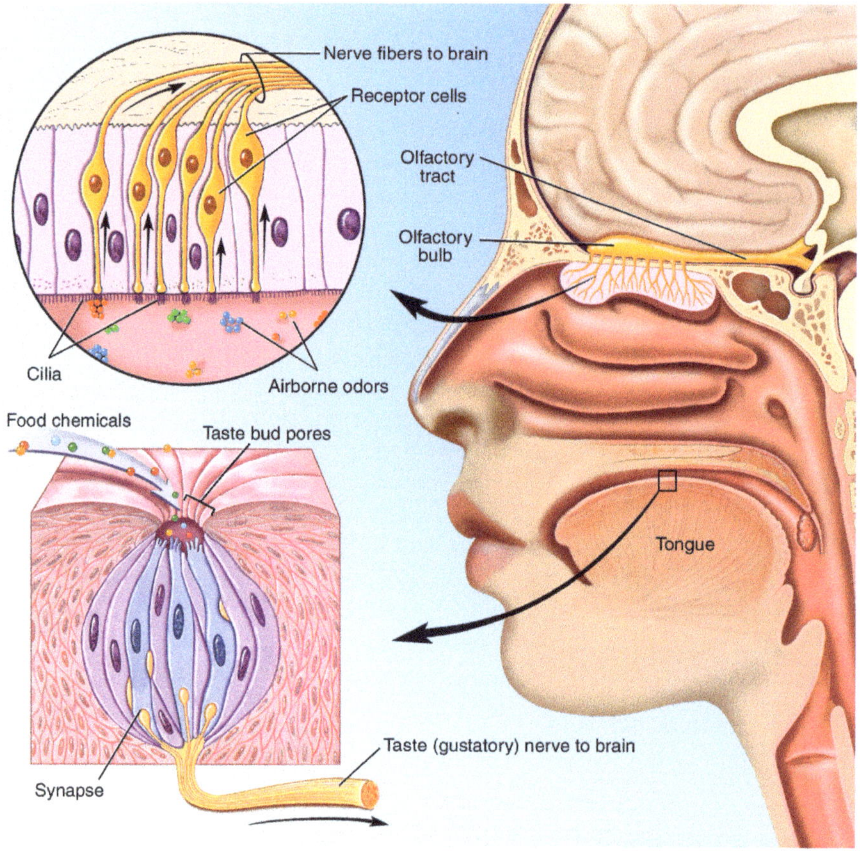

Figure 2.2 Sensory Apparatus for Taste and Smell

It really is all in your head: airborne odorants detected by nasal receptors create signals that are sent to the olfactory bulb, and food chemical tastants detected by taste buds send signals via the gustatory nerve to the thalamus and then the caudal orbital cortex in the brain.

other listed receptors. When transferred to the olfactory bulb and then to the brain, this pattern, "weak, nothing, strong, strong, nothing, strong, nothing, nothing," is interpreted as a highly repulsive smell of oil gone bad. On the other hand, pentanoic acid, which is so similar in structure to heptanoic acid but with one less carbon, does not bind to any of those receptors. Pentanol binds weakly to Receptor B and not to the other receptors listed. For discovering this complicated and unique sensory system and mapping it from the molecular to the cellular level, Axel and Buck were awarded a Nobel Prize in 2004.

Chemical odorant name	Chemical Odorant	Recep A	Recep B	Recep C	Recep D	Recep E	Recep F	Recep G	Recep H
Pentanol		Zero	Low	Zero	Zero	Zero	Zero	Zero	Zero
Pentanoic Acid	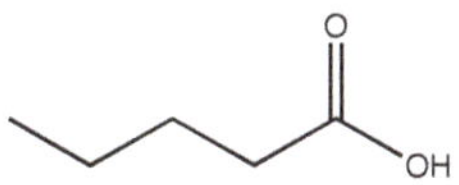	Zero	Zero	Zero	Zero	Zero	Zero	Zero	Zero
Heptanoic acid	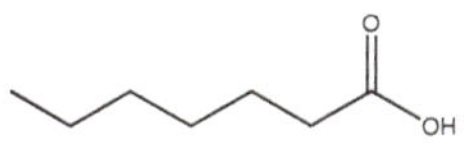	Low	Zero	High	High	Zero	High	Zero	Zero

Figure 2.3 Aroma Molecules Bind to Several of the Hundreds of Different Olfactory Receptors

Odorants create a pattern of neural responses as they bind to different receptors. As an example, consider the three different odorants, pentanol, pentanoic acid, and heptanoic acid, and 8 of the 300 different receptors arbitrarily labeled A through H. For some odorant–receptor pairs, there is no response, such as most of the responses labeled "zero" in the chart. For other odorant–receptor pairs, a weak response is elicited (pentanol with receptor B or heptanoic acid with receptor A). For yet others, odorant–receptor pairs create a strong response (such as heptanoic acid binding to receptors C, D, and F). For odorants, it is not the single receptor it binds to, as with taste, but the pattern of responses from multiple receptors that the brain interprets as a particular smell.

Smelling is extremely important to survival, not only for alerting us to stay away from rotten or spoiled foods, but also to allow us to recognize when there is a threat in our environment, such as a fire or a gas leak.[5] Anosmia, the loss of the sense of smell, can be permanent due to a genetic disorder or temporary due to infection by a virus such as the SARS-CoV-2 that causes COVID-19. The distortion of the sense of smell, called dysosmia, is often associated with disease or treatment regimens such as cancer drugs.

"Gustation" is the scientific term for the sense of taste.

Many master chefs keep a jar full of tasting spoons next to them as they prepare dishes. On cooking shows and podcasts, you will see them repeatedly tasting, adjusting salt, or checking for brightness. What is going on at the molecular level when they taste foods? Food enters the mouth as a mixture of different molecules, some of which are perceived as sweet (candy), salty (peanuts), sour (lemon juice), bitter (coffee), or savory/umami (mushrooms). Some food molecules might be cooling to our tongues (menthol) or create burning sensations (hot peppers). We notice some foods are smooth and creamy (chocolate mousse), and others are thick and dense (oatmeal) or crisp (lettuce). All these perceptions occur because of the responses of particular cells in our mouth to tastants in our food. We will begin by discussing the taste buds found on the tongue and those located on the roof of the mouth and the

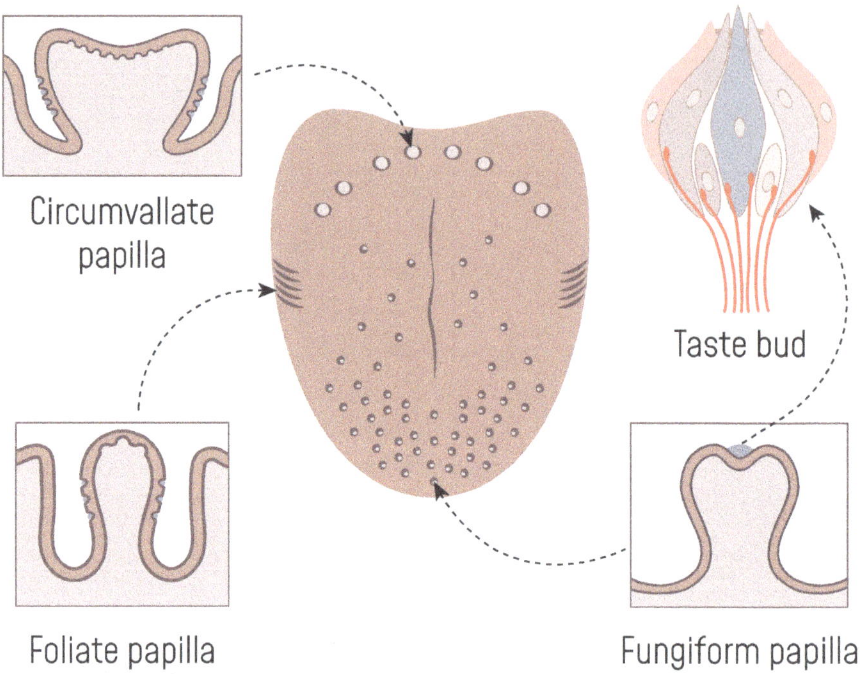

Figure 2.4 Anatomy of the Tongue

Shown are the three types of taste papillae and an expanded image of a taste bud. Each taste bud may contain several different receptor cells (sweet, savory, salt, sour, umami) that coexist and send signals to the cranial nerve.

back of the throat. However, our primary emphasis will be on those found on the tongue.

ANATOMY OF THE TONGUE Your tongue is a muscular organ with as many as 2,000 papillae visible as raised "bumps," as shown in Figure 2.4. About 10,000 taste buds (each containing 50 to 100 taste cells) are found in three types of papillae, fungiform, foliate, and circumvallate, located along the tip, sides, and back of the tongue. The center of the tongue, known as the trigeminal area, is devoid of taste buds. However, it contains filiform papillae that house not taste buds but other sensory cells for temperature, pain, and texture. Together, this sensory apparatus stands ready to send signals from your food to your brain to decide, "YUK" or "YUM." From an evolutionary perspective, the signals allow us to distinguish safe foods from those that might be poisonous or spoiled. At a more complex level, they allow us a certain amount of dietary discretion, perhaps in favor of food that will help us better survive.

Humans vary in the exact number of taste receptors they possess, and receptors change in type and number as we age. It is not true that individuals with an above-average number of taste buds are better tasters, or "supertasters." The typical lifetime of a taste receptor cell is only about two weeks, so we are constantly replacing old cells with new ones, sometimes gaining more of one type or losing others of a different type. The corner slice of frosted birthday cake—the one with the colored frosting roses you coveted as a child—appeals less as our sweet receptors diminish as we age, perhaps replaced by bitter receptors for coffee and chocolate. What you eat can also affect your taste cravings. Certain nutritional deficiencies such as iron, zinc, and A and B vitamins may cause a loss of taste.[6]

Once in the mouth, the food mixes with saliva, and the mixture washes over the tongue, with tastants entering the taste bud through taste pores. Each cell in the bud responds to its target sweet/savory/sour/salty/bitter tastant. Evidence is also mounting for two new types of taste cells: one that responds to fat (oleogustin) and another that responds to starchiness. Once the tastant binds to its cell, we can say that the signal is received. At this point, a few things can happen. Most simply, the stimulated cell can pass the signal to the nerve cells at the base of the taste bud. However, not all taste buds extend down to these nerve cells. Since the cells are also known to secrete small molecules in response to stimulation, it may be that these small molecules communicate with other cells, perhaps causing them to respond in turn. The nerve cell will collect the signals and transfer them to the brain.

The older "map of the tongue" that shows each of the five tastes located in distinct regions of the tongue is overly simplistic. We now know that taste receptors for each type of tastant are found throughout the mouth. Each taste receptor responds only to one type of molecule, but a taste bud contains many different taste receptors and can respond to many tastes. The receptor cells within a taste bud can also interact with one another, attenuating our responses. Try this yourself: Taste a piece of dark chocolate, which tastes bitter, then repeat the taste with a pinch of salt. You will see that the taste has changed significantly, and the bitter taste is "tamed" by the presence of the salt. This feature of cellular cross-talk will become important when we consider flavor pairings later in the chapter.

TASTE RECEPTORS Each class of tastant stimulates its unique receptor cell. In the case of umami, sweet, and bitter substances, this involves a process known as "molecular recognition," in which the receptor protein recognizes its target molecule's shape and charge distribution. The binding of the target molecule by the taste receptor protein produces a cellular change in the receptor, thereby initiating a nerve response.[7]

Each of the taste receptors is a unique protein coded by a unique gene and given names such as "taste 1 receptor" (T1R) or "taste 2 receptor" (T2R). In the case of sour taste cells, the process involves opening a channel protein in the outer membrane of the taste cell through which hydrogen ions generated from acids can selectively pass. This flow of hydrogen ions generates an electrical current, and thus, a nerve signal is created. Salty receptors act similarly, except the signal is a sodium ion. When stimulated, all taste receptors change in ways that depend upon cell type but ultimately lead to a nerve response. Taste receptors are exquisitely sensitive to slight changes in the chemical structures of the tastants they bind. For example, sugars and artificial sweeteners bind to the same sweet receptors and elicit a sweet response in the brain. However, sugars activate an enzyme pathway known as phospholipase C (PLC), while artificial sweeteners activate an enzyme pathway known as adenyl cyclase (AC). These activation pathways are not unique to sweet receptor cells. Upon binding monosodium glutamate, the umami receptors stimulate the PLC pathway. Bitter signal receptors activate both the PLC and the AC pathways.

We have seen that taste cells do not act independently of one another in activating nerve cells. It is also true that taste receptors can interact with one another. Chefs consider these interactions when creating new dishes or improving old ones.

SWEET, UMAMI, AND BITTER RECEPTORS USE MOLECULAR RECOGNITION. Sweet taste receptors (T1R2 and T1R3, as shown in Figure 2.5) create what scientists call "hedonic pleasure" that seems to be baked into our species' need to seek out easily digestible and highly calorific sugars. Taste receptors appear elsewhere in our bodies. These extraoral receptors have been found in the intestines, pancreas, bladder, and brain. These remote receptors may serve a function not related to taste but in monitoring sugar levels that are important in maintaining an adequate energy source for the body.

Umami cells have T1R-type taste receptors that bind to all twenty amino acids (the building blocks of protein), as shown in Figure 2.5. However, the cells show a much more intense response when they bind to the amino acid glutamate or glutamic acid. As food manufacturers know, umami taste cells produce an even greater response when a second molecule known as a nucleotide is also present. We have already seen the nucleotide adenosine in the energy molecule ATP in Chapter 1. Food flavor is enhanced when a glutamate molecule, present as monosodium glutamate, pairs up with adenosine. Nucleotides are abundant in animal muscles, especially shrimp, and organs such as the liver, but we can also obtain them from leafy greens and mushrooms. This feature, the binding of one molecule (here a nucleotide) at one site on the protein, changing the affinity of the protein for a different molecule (here the amino acid), is common in biology and is called allostery.

Humans have at least twenty-five different kinds of bitter receptors (T2Rs, as shown in Figure 2.5) collectively known as our "bitter taste receptor reper-

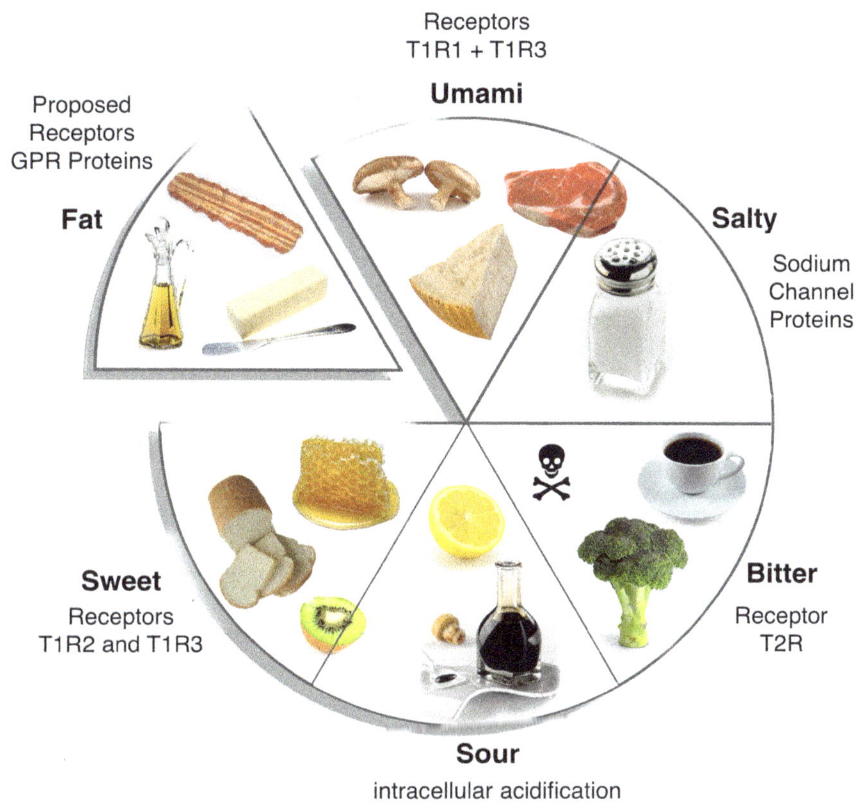

Figure 2.5 Flavor Molecules React with Receptors and Channels in Taste Buds

Receptor molecules for sweet, umami, and bitter (and maybe even fat) bind to their specific tastants, and a nerve response is initiated and sent through an intracellular signal amplification system. Salty and acidic tastants elicit a nerve response when they pass through specific protein channels in the taste bud cells.

toire." Do not feel too special; mice have thirty-five. Some are specific to one molecule type, but others bind to several bitter molecules. Interestingly, the number of molecules recognized as bitter exceeds the number of different types of receptors.

Since most poisonous molecules are bitter, recognizing bitter flavors was essential to early human survival. The distaste often experienced from bitter compounds makes it difficult for some people to eat vegetables or to comply with drug regimens. For example, some extra virgin olive oils contain bitter but very healthy antioxidants—and it takes time to develop a tolerance for these very bitter foods. Exposure to bitter foods at an early age and positive associations with them can build an appreciation for them. A taste for bitter foods such as chocolate, coffee, and even cocktails made with "bitters" often develops as we age. Like sweet receptors, cells that bind bitter molecules exist outside the mouth, in the intestines, brain, bladder, and lower and upper respiratory tract.[8] The precise role of these cells is an area of active research. Still, early results suggest a role in helping prevent respiratory infections and acting as a bronchodilator and playing a role in nonsteroidal treatments for treating asthma. We also know that individuals possess different sets of bitter taste receptors, a characteristic known as genetic polymorphism. As a result, some individuals have difficulty consuming bitter foods such as some vegetables, which are very nutritious but can contain bitter antioxidants. Individuals can be genetically predisposed to be sensitive to bitter foods, as was described in a classic study by Arthur Fox on variations in the bitter taste response in 1932.[9] More than seventy years later, in 2003, it was recognized that this is due to the variations in the TAS2R38 gene for bitter receptors.[10] This genetic variation can affect whether or not you like bitter vegetables and, therefore, your food intake and, ultimately, health outcomes.

SOUR AND SALTY TASTE CELLS HAVE PROTON AND ION CHANNELS. Sour foods like vinegar or lemon contain acidic groups with excess positively charged hydrogen ions (H^+). We will learn much more about acids in Chapter 4 when considering how they can be used to "cook" food. But how do they cause a "sour" taste response? Consumption of sour foods causes a gradient to be created between the outside of the cell (lots of H^+ from the food) and the inside of the cell (much less H^+). Systems become unstable whenever there is an imbalance in concentration or charge, and this creates a driving force to balance the H^+ on both sides of the membrane. Hydrogen ions cannot travel freely through the nonpolar cell membrane because of their charge. Fortunately, channel proteins provide a pathway for the protons to enter the cell in the membrane of sour taste receptors. The flow of hydrogen ions changes the taste cell's overall charge, eventually sending a nerve signal to the brain.

Salty foods contain an abundance of sodium chloride, NaCl, which comprises Na^+ and Cl^- ions. Na^+, like H^+, cannot travel through the membrane independently because of its charge. The salt receptor cells contain channel proteins, which open when the concentration of Na^+ (or some other ions with a +1 charge) outside the cell is greater than the concentration inside the cell. The positive ions of the salt flow from regions of high concentration to regions of lower concentration; this redistribution of charged ions, once again, creates a change in the receptor cell that sends a nerve signal to the brain.

TASTE POTENCY IS NOT LINEAR. Do all sugars taste sweet? Or, more scientifically, is the response of the sweet receptor to all sugar molecules the same? Is the range of responses (the potency) the same for all taste receptors? Table 2–1 shows that the

Table 2–1 **Taste Response Is Not Linear**

Sour	Relative Taste Potency	Bitter	Relative Taste Potency	Sweet	Relative Taste Potency	Salty	Relative Taste Potency
Hydrochloric acid	1	Quinine	1	Sucrose	1	Sodium chloride	1
Acetic acid	0.55	Caffeine	0.4	Aspartame	150	Sodium fluoride	2
Citric Acid	0.46	Nicotine	1.3	Glucose	0.8	Lithium chloride	0.4
Lactic acid	0.85	Denatonium	1,000	Sucronic acid	250,000	Ammonium chloride	2.5

SOURCE: Data from D. Eric Walters, Frank T. Orthoefer, and Grant E. DuBois, eds., *Sweeteners: Discovery, Molecular Design, and Chemoreception*, vol. 450, ACS Symposium Series (Washington, DC: American Chemical Society, 1991), https://doi.org/10.1021/bk-1991–0450.

NOTE: In this table, the response of one tastant, typically the most common one associated with that taste, is assigned the response taste potency (RTP) of 1, and all other tastants for that response are measured relative to that standard. A taste cell does not produce the same response when stimulated by a different but related tastant, such as the bitter tastants quinine and caffeine. Both quinine and caffeine bind to the bitter receptor, but one molecule of caffeine yields a response that is less than half as strong as one molecule of quinine, as illustrated by the RTP of 0.4 for caffeine and 1 for quinine. It is interesting that the ranges of responses for Sour (RTP from 0.46 to 1.0) and Salty (RTP from 0.4 to 2.5) are much narrower than the ranges of responses for Bitter (RTP from 0.4 to 1,000) and Sweet (RTP from 0.8 to 250,000).

taste cells have very different variabilities in the thresholds for taste detection.* Sour and salty taste receptors have a very narrow range of responses. Hydrochloric acid, the sourest molecule tested, is only 2.2 times (1.0/0.46) more sour than citric acid, the least sour molecule tested. For the salty taste response, ammonium chloride is only 6.2 times (2.5/0.4) as salty as lithium chloride—which you can find in trace amounts in cereals and grains. These two taste buds have a similar response mechanism; they are channels for either hydrogen ions (H^+) in the case of the sour receptor or monovalent positive ions ($Na^+/Li^+/NH_4^+$) in the case of the salty receptor.

By contrast, sweet, umami, and bitter receptors have a much greater response range, as highlighted in Figure 2.6. The artificial sweetener sucronic acid is 250,000 times sweeter than glucose, and the synthetic molecule denatonium is 2,500 times more bitter than caffeine.[11] Why do some taste responses show such an extensive range while others do not? The difference is that the channel proteins for sour and salty taste responses can never match the sensitivity of the sweet, bitter, or umami taste responses, which operate by molecular recognition. A physical change occurs in the receptor's shape when a sweet, bitter, or umami tastant binds each to its own receptor type in the membrane. This change causes a change in another molecule and another and another—a process that amplifies

* In these studies, the detectable level for one compound, usually the most common or most typical, is set as 1, and all the other tastants are measured relative to that and measured as relative taste potency, or RTP.

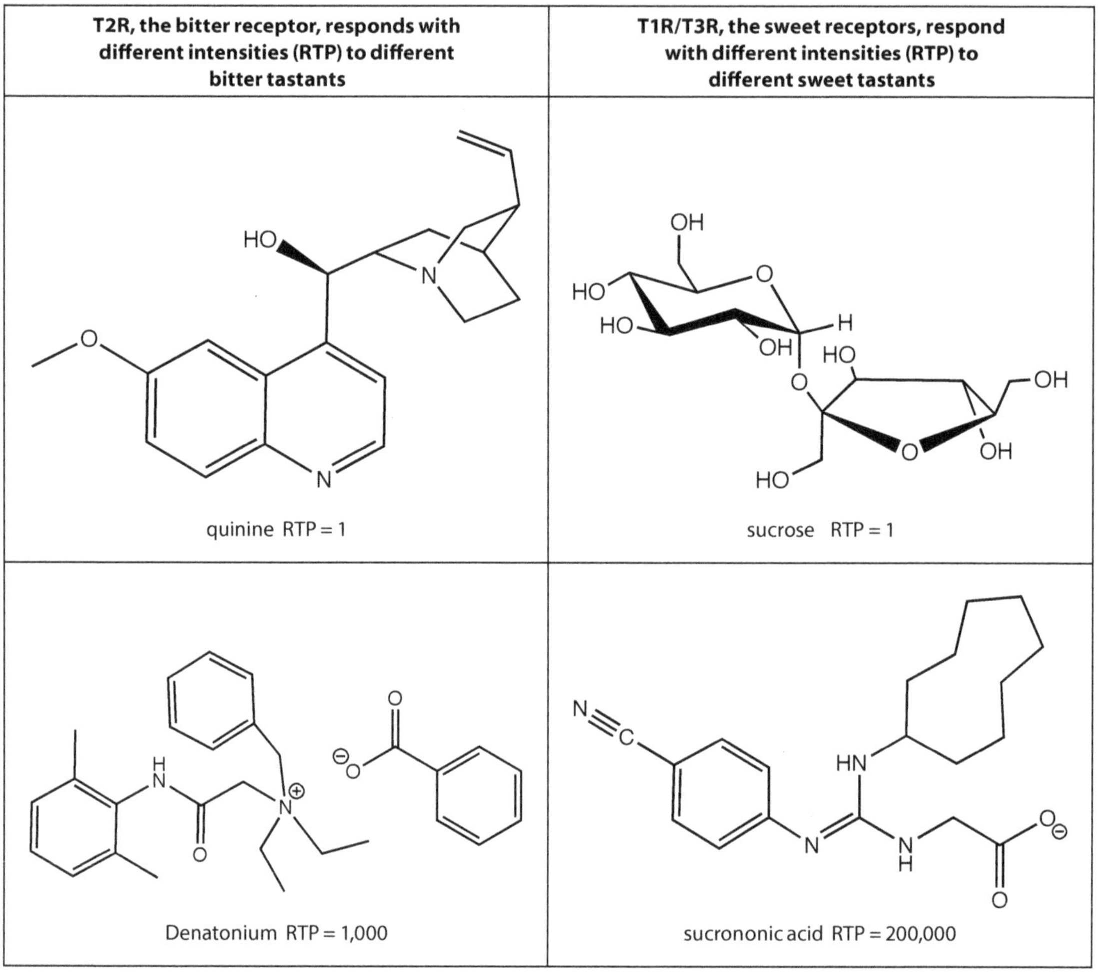

Figure 2.6 Dynamic Range of the Bitter and Sweet Receptors

Quinine and denatonium exhibit a bitter relative taste potency (RTP) of 1 to 1000, while the tastants sucrose and sucrononic acid exhibit an even greater range of sweet relative taste potencies of 1 to 200,000.

the original signal. Sucronic acid and denatonium must bind in such a way to sweet and bitter receptors to better amplify the sweet and bitter signals. The same phenomenon cannot happen in the sour and salty taste responses, where a change in concentration creates the signal as an ion passes through a channel protein. Without a binding event, there is a much narrower range of responses.

Receptors for sweet, bitter, and umami can also be saturated. You can convince yourself of this by letting a hard candy sit on your tongue for several minutes. After a while, despite its presence on your tongue, you can no longer taste the candy's sweetness.

CHEMESTHESIS: TONGUE'S SENSITIVITY TO TOUCH, TEMPERATURE, AND PAIN Biting into a hot chili pepper will cause a burning heat to spread throughout your mouth, even though the food is at room temperature. Your face will turn red, your tongue will burn, and you may begin to sweat. This response is characterized as chemesthesis, a sensation of tingling, buzzing, cooling, or warming, and includes not only the burning from chilies, but the warming from cloves and the cooling from mint. These responses are made possible by a large family of proteins called transient receptor potential proteins (TRP). They are unusual because they act as calcium ion channels *and* sensors for a secondary trigger molecule (such as capsaicin in peppers, cinnamaldehyde in cinnamon, or menthol in mint). In some TRP cells, a signal is sent to your brain when the food in your mouth is too hot or too cold, but in some other cases, the alerts, like false alarms, are sent when a particular trigger molecule is bound, regardless of the temperature of the food. Each member of the TRP family responds to a specific temperature range and a certain type of chemical in food, as shown in Figure 2.7.[12] A model of a filiform papillae cell membrane with the TRP ion channel protein (purple cylinder with a darker purple channel through the middle) is shown in Figure 2.8. No ions flow when the channel is closed in the absence of hot or cold foods or trigger molecules. When we consume food at a specific temperature or food containing trigger molecules, the channel opens, and positively charged ions flow from the outside of the cell to the inside of the cell to create an electrical signal that your brain associates with a particular temperature.

Cinnamaldehyde, shown in Figure 2.9, is a trigger molecule responsible for the spiciness of cinnamon. It binds to a particular type of TRP known as TRPA1 and causes the channel protein to open, allowing ions to flow into the cell. This flow of ions causes a nerve impulse to be sent to the brain and interpreted as "irritant" or "pain." Raw garlic, horseradish, and mustard all contain trigger molecules. We perceive these irritants as more than spicy, maybe even a bit painful; our eyes will often tear up, we may begin to sweat, and we will grab water, milk, or yogurt if it is nearby. Acidic materials such as lemonade or orange juice can help neutralize the spice. Milk and yogurt are slightly acidic, and their high-fat content can dissolve many fat-soluble irritants, helping remove them from the saliva and ease discomfort.[13] By contrast, grabbing for water or beer will not help as these molecules are not very water-soluble, and all the water might do is spread the molecule around your mouth.

Capitalizing on consumers' love of extreme flavors, some candy companies add irritants to their confections to appeal to this audience. One such candy is the Atomic Fireball jawbreaker, made by Ferrara,[14] which contains capsaicin, an irritant found in hot peppers such as jalapeños, as shown in Figure 2.9. It

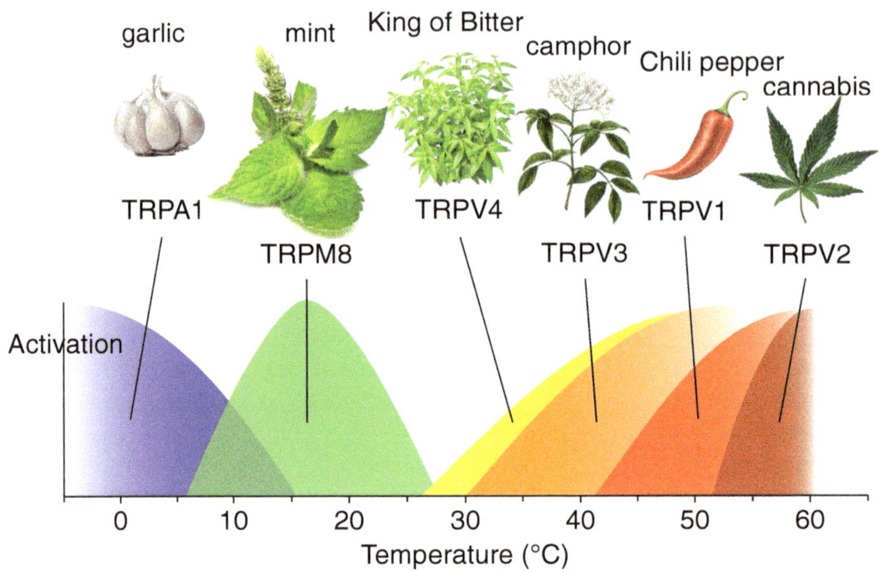

Figure 2.7 Tastants Produce Sensations of Hot and Cold through TRP Proteins

Temperature sensations are created from room-temperature foods that bind to different members of the TRP family of transient receptor potential proteins. Different reactive molecules in these foods bind to the receptors, causing the opening or closing of ion channels in the membrane and creating an electrical signal that our brains associate with a particular temperature.

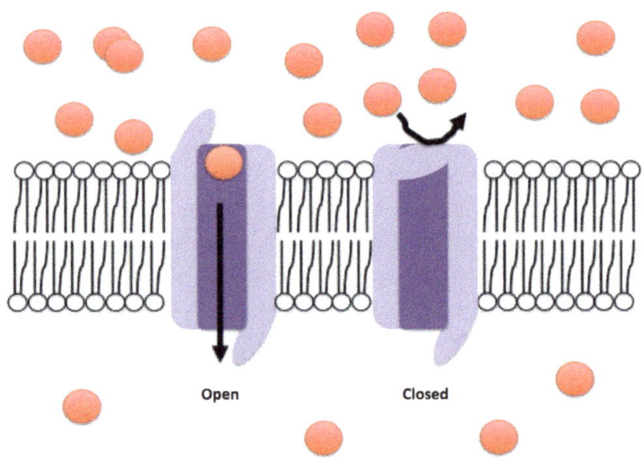

Figure 2.8 Thermal Tastants Cause TRP Ion Channels to Open

Shown here is a portion of a cell membrane (black circles and squiggly lines) that contains two TRP ion channel proteins (shown in purple). The protein channel can open or close, which allows or prevents the passage of positively charged ions or cations (shown by the red balls). When you eat something that is thermally hot (above 50°C) or eat food containing a molecule such as capsaicin from hot peppers, the channel opens, and positively charged ions flow from outside to inside. This passage of ions creates an electrical nerve signal that is sent to the brain that signals pain!

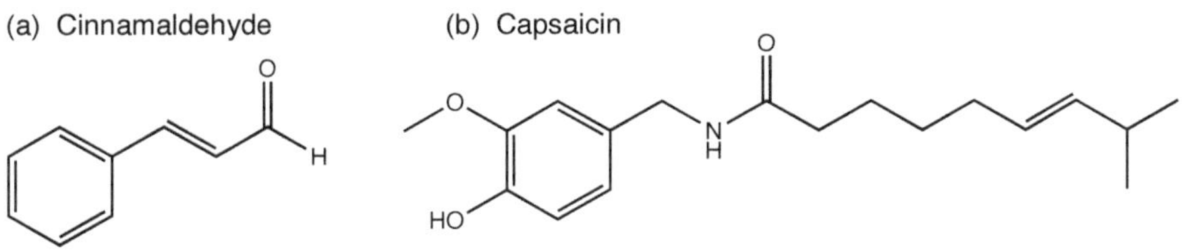

Figure 2.9 Potent Tastants for Spicy HOT

Two examples of food chemicals that trigger TRP channel proteins to create almost painful responses of hotness despite the food being at room temperature are shown. The cinnamaldehyde molecule (a) responsible for the flavor of cinnamon in many baked foods is found in high concentrations in (recently discontinued) Jolly Rancher Cinnamon Fire Candies. The capsaicin molecule (b) is responsible for the flavor of hot peppers and is found in Atomic Fireball candies.

reacts with the TRPV1 channel—and as the candy dissolves, small spikes of pain are sent to your brain—and interpreted by some as pleasurable. Advertising for the Atomic Fireball claims the candy "Burns So Good" and asks, "How much hot can you handle?" Why do some people love capsaicin so much? These compounds are antibacterial and produce sweating, but different motives and reasons, including culture, personality, and gender, no doubt contribute to their popularity.[15]

Why do our tongues have these receptors? The primary role of the TRP receptors in the oral cavity is to check the temperature of your foods or to detect challenging chemicals that might present other challenges to the body. When the food's temperature gets above about 43°C or below 10°C, these receptors will send a STOP signal to the brain—alerting us that this food is too HOT or too COLD. This alarm system prevents us from consuming food that can burn or freeze our mouths and throats. While the chemical irritants are not thermally hot, they are reactive and could cause damage if consumed in large quantities.[16]

Exactly how and why these receptors have also become detectors for other food molecules has yet to be fully understood. However, the 2021 Nobel Prize Committee in Physiology and Medicine recognized the significant advances made in this vital area of research by David Julius. Julius's research uncovered some of the mysterious couplings of these TRP chemical receptors to thermal responses and pain more generally.[17]

The Sense of Sight or Vision

When light particles (photons) scatter off an object and pass through the cornea, they are focused by the lens onto the retina. Each photon is absorbed by a rod or cone cell, and the absorption changes the cell's shape, eventually resulting in a redistribution of calcium ions that creates a nerve impulse. The signals in the optic nerves travel to the brain's visual center, where they are interpreted as an image. Familiar food can trigger a range of emotions, from excitement to disgust, depending on our food memories and personal experiences.

The visual appearance of food, especially how it is plated, can dramatically alter our perceptions of the foods we eat. Plate color, plate shape, neatness of the food, lighting, garnishes, and utensils are just some of the variables that can make a dish taste as good as it looks.[18] Food plating is essential when the goal is to stimulate appetite and increase a consumer's enjoyment. Of course, one expects meticulous food plating from high-end restaurants, but careful food plating can even boost the appetite of Alzheimer's patients.

The Sense of Hearing or Aural Sensitivity

Eating different foods produces sounds of varying frequencies, measured in Hertz (Hz or cycles/second). For instance, crunching on raw carrots produces a frequency of about 1,000 to 2,000 Hz, whereas eating crispy flatbread produces a higher frequency sound of 5,000 Hz. For comparison, a man's deep voice tends to have a much lower frequency, about 130 Hz, and the sound of a crying baby is around 3,500 Hz. Recent research has shown that high-frequency sounds enhance sweetness, while low-frequency sounds bring out bitterness.[19]

When the food is a potato chip, it is all about the crunch or lack thereof. We determine our chips' freshness or staleness based on the crunch sound's level and frequency. Our sense of hearing tells us that potato chips with the loudest crunch are fresh, and those that do not produce the audible crunch are stale. Our brains are particularly attuned to associate the high-frequency crunches of 2,000 to 20,000 Hz with crispier and fresher chips. According to human taste testers, a chip was perceived as crispy when high-frequency crunching sounds were played through headphones, but the same chip was rated as stale and soggy when low-frequency sounds were played.[20]

The Sense of Touch

What foods do you find the most revolting? The most common attribute that unites most foods considered disgusting is their texture. Food with a rubbery or slimy texture is often described as revolting. We must, however, keep in mind that what one culture considers unappetizing, another culture may highly value. There is no doubt that texture is critical in our perception of food.

The "mouthfeel" is determined by many properties, such as elasticity/stiffness, moisture/dryness, and creaminess/roughness. The 2021 Nobel Prize in Physiology and Medicine was awarded to Ardem Patapoutian for his research that led to discovering the mechanism of mechanical sensory perception, or proprioception.[21] The apparatus for proprioception is located on the trigeminal nerve in the center of the tongue. Signals from the trigeminal nerve are eventually integrated into the sensory center of the brain. Many receptors detect pressure, all with a shallow threshold that can rapidly see small changes. These receptors detect how much a food compresses or stretches the receptor, and this change in the pressure on the receptor is communicated to the brain and interpreted as a texture.[22]

A food's texture evolves as it is chewed or masticated; that is, its physical properties change as it breaks into smaller pieces. Before being eventually swallowed, food mixes with saliva, which both hydrates the food and contains thirty different enzymes that begin the digestion of the macronutrients in the food.[23] We will learn much more about this in Chapter 4. Food sensory scientists report that food consumed with your hands is perceived as more enjoyable, filling, and satisfying. One reason may be that the hardness, softness, or creaminess you feel with your hands gives your mouth a preview of what is coming.[24]

SECTION 3: FOOD PAIRING

We have seen that input from all five senses is used in our appreciation of food. How can we use this information to improve the flavor of foods and create new dishes? Food pairing techniques intentionally combine complementary aromas or harmonize dissimilar elements to balance a dish. Flavor pairing theory was introduced in the 1990s by Master Chef Heston Blumenthal, owner of the world-famous Fat Duck restaurant in England. The theory works on the principle that foods that contain some or many of the same aroma-causing molecules will pair or complement each other. These ideas led Blumenthal to create incredibly successful and unusual pairings such as caviar and chocolate.

Some chefs have such finely developed senses of taste and smell that they can do this intuitively. For others, a rational approach can be helpful. Databases, or libraries, have been created to index the tens of thousands of volatile aromas in different foods.[25] Successful and unique food pairings can be created by identifying the major odorants in foods and seeking out those with the same or similar aromatic compounds. Today, resources such as James Briscione's book, *The Flavor Matrix*,[26] help cooks identify complementary and balanced food ingredients. Figure 2.10 integrates Briscione's observations with the biochemical information relating to the types of taste cells to which these tastants bind. Complementary tastes such as chocolate (bitter T2R receptor) and salt (Na^+ channel), or umami (T1R receptors) and fat (GPR receptors), work in harmony (blue boxes in the table), with one tastant enhancing the taste of the

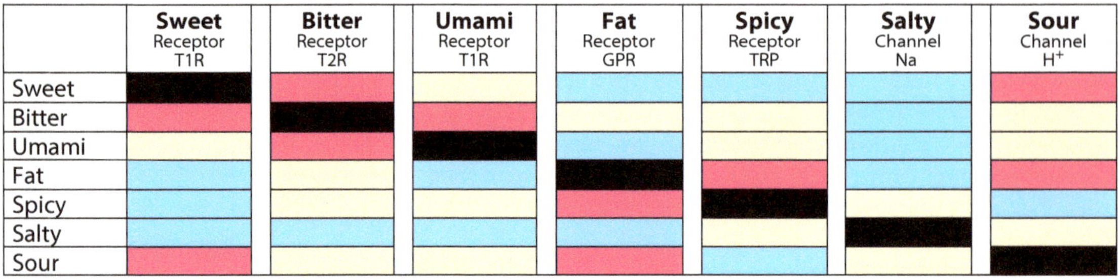

Figure 2.10 Flavor Interaction Matrix Can Guide Food Pairings

Seven taste responses are listed, five of which interact with flavors or tastants (sweet, bitter, umami, fat, and spicy) through receptor proteins (T1R, T2R, GPR, and TRP) and two of which interact with their tastants (salty and sour) through channel proteins for Na^+ or H^+. When a taste cell is stimulated, it can communicate its response to other taste receptors. Complementary flavor pairings (blue boxes) occur when two tastants have similar taste responses, such as umami and salty, and stimulating both cells attenuates the flavor. Balancing interactions (red boxes) occur when two tastants stimulate dissimilar taste responses, such as sweet and sour, and stimulating both creates a different response. Neutral interactions (tan boxes), such as salty and sour, occur when the receptors do not influence each other, and though foods containing these flavors are safe to combine, there is no complement or balance in the pairing. Black boxes on the diagonal just represent the same receptor-taste feature, such as sweet-sweet.

other. Chemical anharmonicity, or the matching of dissimilar tastes to create balance (red boxes in the table), allows us to understand some classic pairings, such as sweet (T1R receptors) and sour (H^+ channel) or bitter (T2R receptors) and sweet (T1R receptors). Some receptors do not seem to talk to one another, as evidenced by the tan boxes in Figure 2.10.

Vast online databases of food components, such as those sourced on the website Foodpairing, allow users to create new pairings by suggesting matched ingredients once a starting food is selected.[27] The program can do this by accessing the databases of flavors, fragrances (aromas), and indices of texture. Once a base food is chosen, the program will search its database of foods to find another food with a complementary aroma profile and contrasting taste and texture profiles. In a trial run, the Foodpairing computer program suggested pairing white asparagus with sweet red cherries. Another successful trial led to a pairing of carrots, instant espresso, and roasted coconut. Delicious!

SUMMARY OF CHAPTER 2

When we eat, the primary sensory input is derived from taste and smell. These sensory systems are based on the interaction of either a tastant or an odorant with a receptor. Taste happens when food molecules in our mouth interact with one type of taste bud (sweet, sour, bitter, salty, and umami). Smell happens when volatile food molecules diffuse into our nasal cavity and bind to one or more of the several hundred different types of receptors, creating graduated responses depending on the receptor. In addition, our food's texture and spiciness and even the sounds generated as we take bites and chew are recognized by other sensory systems, many of which have elements in common with taste and smell. Vision cannot be ignored, as the appearance of food can significantly affect whether it stimulates the appetite. As shown in this chapter, many aspects of this fivefold flavor architecture can be explained using chemical and biochemical principles of molecular recognition, allostery, signal amplification, signal saturation, and harmonies or dissonance of food pairing. All contribute to our understanding of how we taste our foods.

In Part II, we will build on what we learned here regarding the makeup of food and its role in sustaining us (Chapter 1) and the various senses that contribute to our enjoyment of food (Chapter 2). We will delve deeper into the physical and chemical changes that occur at a cellular and molecular level as we transform ingredients into delicious dishes.

PART

II

TRANSFORMING

In Part II, we begin to consider how the tools of the culinary world are, in fact, manifestations of chemical processes. We discover that at a most fundamental level cooking is a series of transformations of molecular structures. Following recipes, we heat, cool, oxidize, acidify, ferment, or mechanically alter ingredients. Oxygen in the air reacts with food molecules, causing browning, caramelization, searing, or spoilage. Knowing what is happening at the molecular level allows us to take actions that minimize, maximize, slow down, or speed up these transformations. Culinary techniques as simple as grilling or as sophisticated as sous vide or ceviché are examples of the underlying chemistry and the transformations of molecular structures that happen with changes in temperature, pressure, or solubility.

In **Chapter 3—Cooking with Fire and Ice**, we explore changes in the molecular structures of food as our raw ingredients (reactants) become plated food (products). Weak bonds can break when the system's energy changes, such as during heating or cooling. Other ways of changing the energy, such as microwave radiation, can also cause transformations. The materials and construction of cookware, the tools for cooking, need to be matched to the technique used for best results. When energy is added to the complex proteins, sugars, and fats, each reacts in unique ways, leading to extraordinary transformations that result in the flavors and aromas that contribute to the enjoyment of a delicious meal.

In **Chapter 4—Transforming Food Using pH, Pressure, or Microbes**, we see that using acids, bases, pressure changes, and fermentation are additional ways to prepare our food with ancient and modern interpretations and popularity. Citrus juices and vinegar can supply an abundance of acidic hydrogen ions, and lye or baking soda can provide a lot of alkaline hydroxide ions, each of which can cause protein unfolding that works for cooking food while maintaining moisture. When considering the sensitivity of cooking transformations to atmospheric pressure, we can make better decisions about time and temperature control, especially for baking. We can get desirable results in the kitchen by artificially increasing or decreasing the pressure. Microbial fermentation might be one of the most ancient food transformation and preservation processes. It is often naturally accompanied by shifts in pH or the breaking down of cell structures to make food more digestible.

In **Chapter 5—Manipulating Texture**, we explore how vital mouthfeel is to our enjoyment of food. Using chemical cross-linking to create natural or artificial polymers, we can encapsulate or thicken food. Emulsions and foams are created by mixing two substances, often of different phases, to create a product in a third phase, for example, egg whites plus air to make meringue. We can develop new phases and stabilize inherently unstable emulsions.

CHAPTER

3

Cooking with Fire and Ice

In this chapter, one of the most common forms of cooking—heating—is explained in terms of the physical and chemical transformations described in **Section 1**. In prehistoric times, cooking happened over a campfire as an exchange of heat between fire and food. Eventually, humans developed cookware that would enable them to manipulate moisture, and in the process they created new cooking techniques such as stewing and braising. Today cooking methods have expanded to include heat flowing from new energy sources to the food via conduction, convection, and radiation. For each dish we prepare, our cookware should be compatible with the heating method, a consideration that will often have a molecular component. In nontraditional low-temperature cooking, heat flows in the opposite direction: from the food to the chilling element, usually a cryogen or cold substance such as liquid nitrogen. Again, it is essential to match the molecular properties of the cookware with the low-temperature methods to be sure we have the most appropriate match that considers heat transfer, durability, and strength. At a cellular level, in addition to the macronutrients in our food, the insides of plant and animal cells are predominantly water. A cell wall (or two in the case of plants) surrounds the internal contents, holding them intact and controlling the passage of material into and out of the cell.

When we physically break cell walls by chopping or chewing, materials within the cell that are kept separate in their chambers can mix, leading to new chemical reactions that begin to transform them. **Section 2** describes transformations in food that happen as the result of the disruption of covalent and non-covalent interactions and as oxygen from the air oxidizes cell components. In **Section 3**, we explore how the molecular structure of each of the macronutrient groups changes with heat and how these changes affect flavor. Finally, in **Section 4**, we explore some of the more complex chemical reactions. Browning reactions can happen enzymatically at room temperature and by heating sugars, causing caramelization. The Maillard reactions between amino acids and purine molecules create delicious flavors. Step-by-step experiments guide you in examining the transformations that happen as you cook an egg and cryo-cook at low temperatures, which can transform food in unique ways.

Two lab experiments based on these techniques can be found in Part IV—Performing.

SECTION 1: COOKING AS A PHYSICAL AND CHEMICAL TRANSFORMATION

At a fundamental level, cooking food is necessary to make foods safe by destroying pathogens that might make us ill. Cooking also begins cellular degradation processes that break down cell walls and cellular fibers that would otherwise make the food too tough to chew or swallow. In that way, cooking or adding heat helps to facilitate digestion along with the mechanical disruption of chewing and the chemical degradation made possible by enzymes. Along the way, cooking develops new flavors and fragrances. Heat can transform food through physical changes that change the physical state of the food (such as the melting of butter) and chemical changes, such as the browning of butter, where the breaking and rearrangement of chemical bonds create new materials. Over time, unique cooking styles or cuisines have evolved in different geographic regions or among people of different ethnic or religious backgrounds. However, they share the same principles of transformation discussed here.

The first step in making food digestible is often to chop, slice, and dice it into smaller "bite-size" pieces. The sharp edges of a knife begin the shearing process, breaking down cell walls and liberating the cell contents. As the cell contents mix, biomolecules that would have been kept separate from each other inside the cell start to mix, causing enzymatic breakdown.

Tears that you shed as you cut an onion provide evidence for this. The molecule that makes you cry, thiopropanethial-s-oxide (TPTO), is created when the cutting process mixes amino acid derivatives from one part of the plant cell with the enzyme known as allinase from another part of the cell. A chemist will say that TPTO is volatile, which means that some of the molecules can evaporate from the onion, enter the gas phase, and diffuse through the air until it reaches your eyes. Once there, TPTO binds to specific tissues and irritates them, causing tears as your body tries to wash the irritant away.

For some foods, the next step is heating. Heating plant and animal cells has various effects. One is boiling the intracellular water, which creates pressure as the liquid water expands into steam, eventually causing cell walls to burst. Adding heat also causes cell membrane components to become disordered, melting down the walls that hold the cell together and spilling out the cell's contents. With the walls down, again, the macronutrients and micronutrients in the cell are made available to your body. Enzymes from your saliva begin breaking down the food molecules into their component building blocks that are useful for energy or growth.

A Primer on Heat Transfer Processes in Cooking

Scientists love to argue about the nature of heat. Is it an actual thing like light or electricity or a process like freezing or boiling? For our purposes here, we will use the concept that heat is a measure of the thermal energy in a system as measured by its temperature. Thermal energy is the energy a system has because of the jiggling, twisting, and random motion of the atoms and molecules. Matter has three distinct phases, solid, liquid, and gas, each with its characteristic degree of motion. In the solid phase, atoms and molecules stay put, though they may stretch their bonds or vibrate some. Atoms of a solid will sometimes have strong interactions because they are close to one another. So it will take energy to break those interactions and cause the solid to turn into a liquid, a phase transition called melting. In liquids, there is more motion as atoms and molecules slip and slide over each other, but it still takes energy to get that water to become gaseous, another phase transition known as boiling. In the gas phase, molecules move with constant random motion, with little interaction between the particles. The boiling water or melting butter temperature remains constant during these phase transitions as the added energy is used to change the phase rather than raise the temperature as long as the pressure is held constant. One example of a cooking technique that takes advantage of the constant temperature heat provided by boiling water is a double boiler, or bain-marie, in which a food item, such as chocolate, is melted in a smaller pot and held just on top of the boiling water.

In phase transitions, the state of the material (solid, liquid, or gas) changes, but chemical bonds remain intact. Heating foods above their phase transition temperatures can do more than change the phase; it can break or form chemical bonds. Cooking involves adding energy through conduction, convection, or

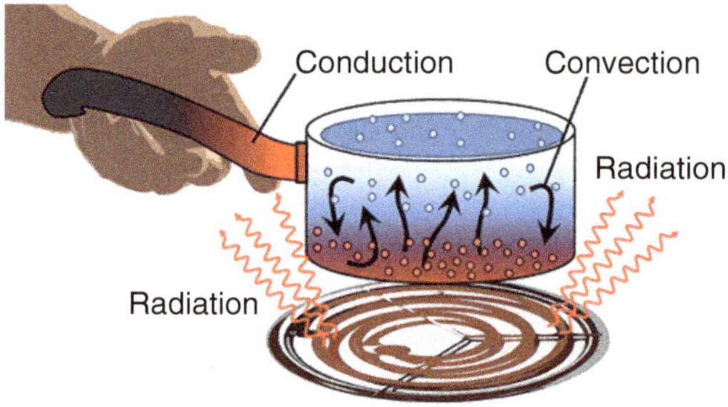

Figure 3.1 Three Mechanisms for Transferring Heat While Cooking

Heat is transferred to our food directly by conduction indirectly, through a medium such as water or air by convection, and through space by radiation.

radiation, as shown in Figure 3.1 and summarized in Table 3–1. Understanding the different types of heat transfer, the reasons for preferring one heating method over another, and the importance of selecting appropriate cookware is essential to mastering the food transformation process. This knowledge will help you make informed decisions when cooking.

Transfer of Heat by Conduction

In conduction, heat transfer occurs through direct contact from a hot to a cold solid. A fundamental law of physics, called the second law of thermodynamics, says that spontaneous heat flow *always* occurs from a material at a high temperature to a material at a low temperature, never the other way. How efficiently heat transfers depends on the atomic or molecular structure of the two materials and how easy it is to transfer that thermal motion from one material to another. The molecules of certain solids, such as glass, ceramic, or wood, are rigid and resist movement and are called thermal insulators. Other materials, such as copper and aluminum, are made up of molecules whose motion is less rigid, and cookware made from these materials must be good conductors of heat. You may have had the experience of grabbing a metal spoon that has been sitting in a pot of hot soup—ouch! The heat gets transferred to your hand. You would have been better off to have chosen a wooden spoon. When it comes to cookware, much trial and error goes into manufacturing pots and pans with the precise layering of materials to make a product that is durable, able to transfer heat efficiently, and has a nonstick surface that is safe to use. In making cookware, the properties of pure substances, elements, and mixtures are

Table 3–1 **The Variety of Cooking Techniques Organized by Type of Heat Transfer**

Cooking Techniques

Dry	*Conduction*	Dry roasting, hot salt frying, searing
	Convection	Baking, roasting, smoking
	Radiation	Grilling, roasting, rotisserie, toasting
Wet	*High heat*	Blanching, boiling, parboiling, shocking
	Low heat	Coddling, creaming, infusion, poaching, simmering, slow cooker, smothering, sous vide, steeping, stewing
	Indirect heat	Bain-marie, double boiling, double steaming, steaming
Fat-based	*High heat*	Blackening, browning, deep frying, pan-frying, reduction, shallow frying, stir frying (*bao*), sautéing
	Low heat	Gentle frying, sweating
Mixed medium		Barbecuing, braising, flambé, fricassee, indirect grilling, plank cooking, stir frying (*chao*)
Device-based		Air frying, microwaving, pressure cooking, pressure frying
Nonheat		Curing, fermenting, pickling, scouring

SOURCE: Adapted from Wikipedia, https://en.wikipedia.org/wiki/Cooking.

important. Figure 3.2, Periodic Table of Cookware, highlights the elemental composition of most cookware used with conductive heating.*

Copper cookware is beautiful, heats evenly, and is quick to heat and cool, which makes it a nimble cooking tool for delicate foods. It is a favorite for many chefs despite its high cost. Figure 3.2 shows that copper cookware, highlighted in red, is made from copper, Cu, atomic number 29. Cu is a metal with a high melting temperature of 1,084°C and is ductile, meaning it can be formed into many shapes (coins and roofing material, as well as cookware). Metal pots dating from 250 BCE show that metalsmiths have long known that hammering Cu strengthens the metal. Even so, Cu cookware deforms easily, and the surface is easily scratched or dented, so pots thicker than 2.5 mm are recommended. Cu also tarnishes, so some upkeep is required. Cu conducts electricity and reacts with acid and alkaline foods, giving food a metallic taste. For this reason, Cu cookware is usually lined with the less reactive metal, tin, Sn, atomic number

* Quick review: The Periodic Table organizes all known pure substances, or elements, in rows (called periods) and columns (called groups). Each element is assigned an atomic number, which is the number of protons, or particles containing a positive charge, in the nucleus (central core) of an atom of that element. The Periodic Table orders elements in terms of increasing atomic number. Positive charges in the nucleus of an element are balanced by negatively charged electrons that surround the nucleus and make the element neutral. Electrons are distributed into different shells each of which has a particular occupancy limit. When the shell is filled, the next element starts a new row or period. The first element in each row has one electron in a new shell called the valence shell, and the last element in the row has filled the shell. Only electrons in the valence shell participate in chemical reactions.

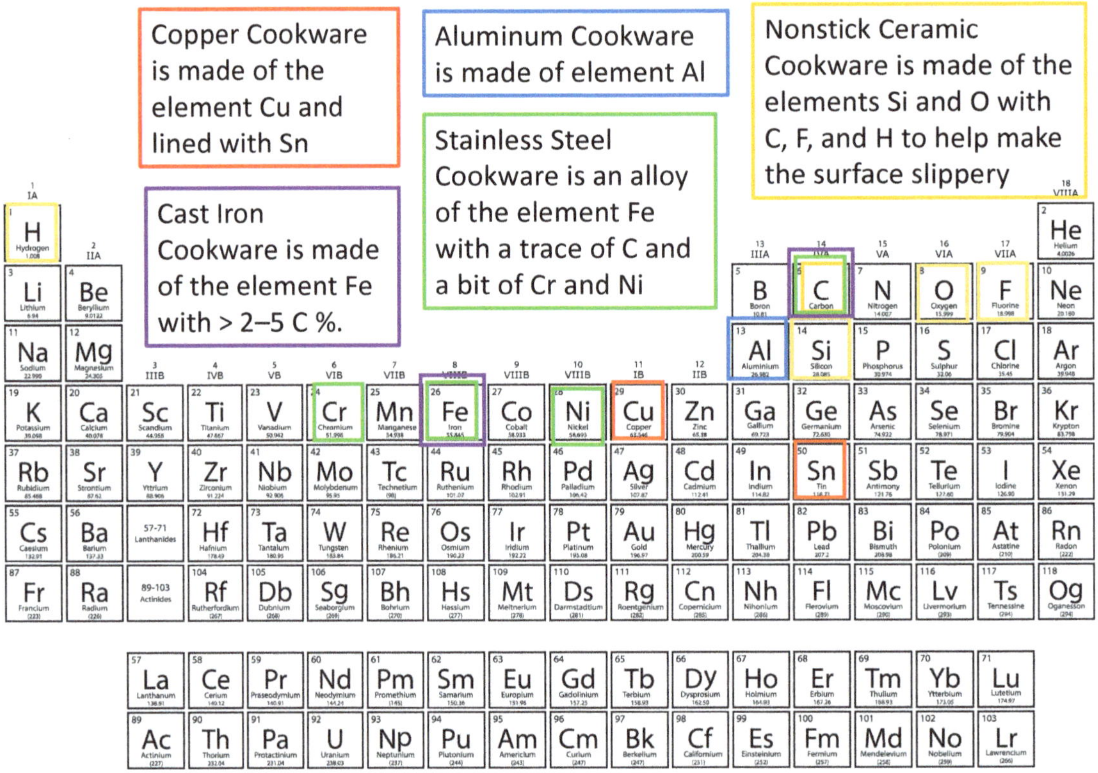

Figure 3.2 Periodic Table of Cookware

There are five commonly used types of cookware: copper (red), aluminum (blue), stainless steel (green), cast iron (purple), and nonstick ceramic cookware (yellow). Using the same color code, elements that make up each type of cookware are surrounded by colored boxes in the Periodic Table.

50. Sn melts at 231°C, so never leave the cookware empty on a hot burner, or your expensive pot will be destroyed. The lining also may wear thin over time, and the cookware may need to be retinned.

Aluminum cookware, as highlighted in blue in Figure 3.2, is made with aluminum, Al, atomic number 13. It is great at heat distribution but has the downside of being reactive. Over time, Al leaches into food, especially acidic food such as tomato sauce, and gives the food a grayish cast and a metallic taste. Al pans can also dent or warp, which prevents the pan from maintaining good contact with the cooking surface, a problem for lightweight, inexpensive pans manufactured by stamping. Forged Al pans are a bit higher in quality and stronger. In die casting, the most expensive manufacturing method, Al is melted at 660°C and forced under high pressure into a mold with the correct shape. The product is heavy and holds its shape well. Some Al pans undergo an electrochemical process known as hard anodization, which strengthens and darkens the metal, protecting it from denting and preventing it from interacting with food.

Steel is an alloy or mixture of iron, Fe, atomic number 26, and a trace amount of carbon, C, atomic number 6. It is prized for its beautiful appearance and strength but is not a good conductor of heat, and its lustrous appearance is hard to maintain. Cookware is most often made from stainless steel, highlighted in green in Figure 3.2, which is produced by adding chromium, Cr, atomic number 24, to the steel alloy to prevent corrosion and oxidation. More expensive alloys add nickel, Ni, atomic number 28, to increase their strength and durability. Stainless cookware is often described with a ratio of two numbers—such as 18/10, 18/8, or 18/0—the ratio of the weight % of Cr to Ni that has been added to the steel alloy. Stainless cookware is often clad, meaning it has alternating layers of stainless steel and other metals such as Al or steel. An interior core of Al ensures even heating. Disadvantages are that some foods, especially eggs, stick to stainless steel cookware, though pans can be easily cleaned with soap and water or a nonabrasive cleaner.

Cast iron is another Fe and C alloy, highlighted in purple in Figure 3.2. At 2%, it has a higher C content than steel, which makes it brittle, extremely hard, and dense. It is impossible to forge it into shape, so it must be heated to 1,204°C and then poured into casts or molds when molten, hence *cast iron*. A cast Fe pan lasts virtually forever; the first known cast Fe pots were used in China in 220 CE. Because it is so dense and has a high heat capacity,† cast Fe retains the warmth and resists changes in temperature when cold foods are added. Cast Fe's higher density makes it a powerhouse of high thermal energy, which is why it can sear steak effectively. Cast Fe may be treated with a coating by the manufacturer, such as porcelain enamel, or it may be produced uncoated. Uncoated cast iron requires a seasoning with oil before use to create a nonstick surface, while coated cast iron usually requires no seasoning.

Nonstick cookware, highlighted in yellow in Figure 3.2, is often made by coating the surface of Al or stainless cookware with materials that create a slippery surface that makes cooking foods such as eggs extremely easy and minimizes cleanup. Unfortunately, none of these coatings lasts forever, so the best advice is not to invest lots of money in nonstick cookware. Three basic classes of materials can be used to provide a nonstick surface.

- Synthetic polymers such as Teflon (polytetrafluoroethylene, or PTFE) create extremely nonadhesive surfaces. Chefs using any polymeric nonstick surface must avoid heating the cookware without food as the polymers can degrade at high temperatures and produce toxic by-products. Scouring pads and metal spatulas pit the surface and should be avoided.

† Heat capacity measures how much energy it takes to change a substance's temperature by one degree Celsius.

- Ceramic coatings provide a nonstick surface that offers an ecological alternative to PTFE, is tolerant of high temperatures, and is scratch-resistant. Some ceramic surfaces are infused with diamond dust to increase durability and thermal conductivity. Diamond, a form of carbon, C, is one of the hardest materials known and has a thermal conductivity that is 5 times greater than Cu. Oil-infused ceramic coatings eliminate the need to add oil or butter for health-conscious cooking and convenience.
- Enamel coatings are made by fusing powdered glass (silicon dioxide, or SiO_2) onto steel or some other core cookware substrate. Pigments can be added before firing at 850°C, and the resultant cookware can be beautiful. Covering steel elements with enamel protects the base material from rust and guarantees health and safety for consumers. Enamel coatings also make cookware durable and dishwasher safe.

Transfer of Heat by Convection

Convective cooking is a process by which heat transfer occurs through contact with hot fluids, such as air, water, or oil. Convection stoves circulate hot air through the stove—keeping temperatures in the oven constant and even. Stovetop cooking methods, such as those with water (steaming, boiling, poaching), air (air frying, popcorn popper), and oil (sautéing, deep fat frying, stir-frying), all involve convective processes. This heating method is slower because the fluid needs to be heated to the correct temperature first and then carefully adjusted to avoid overheating the food. However, the liquids, especially oil, can provide even heating and impart additional flavor to the food.

Transfer of Heat by Radiation

Radiative cooking involves the transfer of heat using energy in the form of electromagnetic waves.[1,2] Cooking with radiation consists of using various types of radiation, including microwave, infrared, and, less frequently, ultraviolet or gamma radiation. Unlike water or sound waves, electromagnetic radiation can travel through a vacuum; it does not need air or water to get from one place to another. That is why the light of a star can get from a point millions of miles away and travel through the vacuum of space to reach Earth. The speed of the light that travels through space is super-fast—300,000,000 meters/second (m/s), or 3×10^8 m/s. For comparison, in 2009, the Jamaican sprinter Usain Bolt ran the 100 m dash in 9.8 s, for a speed of 10.4 m/s. Light travels roughly 10 million times faster than that.

What about the other properties of electromagnetic radiation? As the name suggests, all electromagnetic radiation has electric and magnetic fields. It is impossible to separate these two characteristics and still have electromagnetic radiation. Both the electric and magnetic components have a measurable

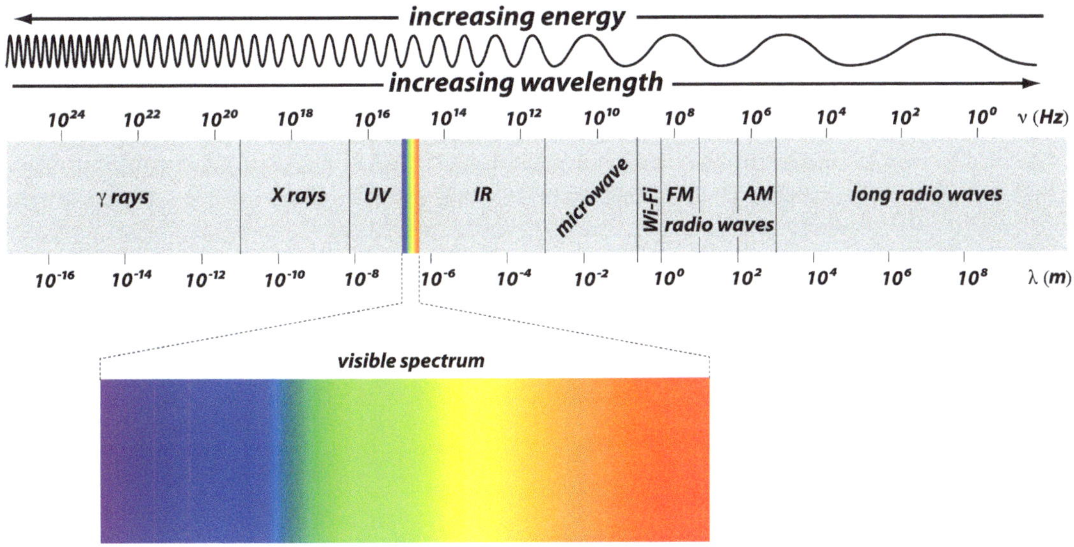

Figure 3.3 Electromagnetic Spectrum

This image shows the ranges of wavelength (10^{-9} nm–10^{3}m), frequency (10^{24} Hz–10^{2} Hz), and names of the regions of the spectrum (gamma rays, x-rays, ultraviolet rays, visible light, infrared, microwaves, radio waves). The narrow part of electromagnetic radiation that we can see, called visible light, is expanded in the image. Note that the the wavelength decreases as the energy and frequency increase.

field strength, and if you were to stand in one place as the wave passed, you could detect the strength of the fields oscillate; that is, the field strength smoothly increases, then decreases, then increases again, on and on until the wave has passed by. You can see that a water wave observed from the shore has an amplitude that increases, decreases, and then increases again. We can measure that oscillation in cycles per second, or Hertz (Hz), which we also call the frequency. Using the analogy of a water wave, you might notice, for example, that there are 5 seconds between peaks of the wave, which means that the wave has a frequency of 1 cycle every 5 seconds, or 0.20 Hz. Electromagnetic radiation varies in frequency from 10^{24} Hz for gamma rays to 10^{2} Hz for radio waves.

Electromagnetic radiation has another property called wavelength, which refers to the distance between two adjacent peaks in the wave. The wavelength is greater for lower-frequency radiation. Figure 3.3 is a graph called the electromagnetic spectrum, where we can see that different wavelengths of light are associated with different types of radiation. Visible light, shown by the expanded colored excerpt of the grid at the bottom of the figure, is perhaps the most familiar form. However, it is just a tiny slice of the wide range of electromagnetic radiation. Of course, chefs rely daily on the absorbance and transmittance of visible light when choosing ingredients and using their eyes to distinguish a ripe red tomato from a green tomato and to determine the composition of a well-plated dish.

Many types of radiation have unique applications in the preparation of food. Restaurants use infrared radiation (IR) from hot lights to keep food warm (wavelength ~0.000001 m, or 10^{-6} m), and food processing plants use gamma radiation to sterilize foods for extended storage (wavelength ~10^{-12} m). It is essential to understand that **using electromagnetic waves to irradiate food does not make it radioactive**. During irradiation, microwaves pass through the foods, causing heating. X-rays and gamma rays (γ-rays) also pass through food, destroying or inactivating bacteria and viruses that cause foodborne illness. In all cases, the radiation is gone once removed from the power source, like pulling the plug turns off a light.

Microwave ovens use electromagnetic radiation with a wavelength of about 12.0 cm or a frequency of 2.45 gigahertz (GHz) (2.45×10^9 Hz).[3] The microwaves are produced inside the oven by a magnetron electron tube. They travel through glass, plastic, paper, or other similar materials. The water molecules in the food absorb the microwaves, causing them to rotate more vigorously—thus increasing their thermal energy.[4] Foods with high water content, such as popcorn, respond well to heating in a microwave; those with low water content are less responsive. Metal reflects microwaves, so a fine metal mesh on the oven's glass door allows you to see through to the heated food but prevents the microwaves from escaping. Inside the oven, metal objects will obstruct the passage of the microwaves and cause a buildup of charge, which can spark and pop and eventually ignite, which is why metal containers and aluminum foil should never be used in the microwave. With the power to the oven off, the microwaves disappear, so one does not have to worry about residual radiation in the food. Foods cooked in the microwave tend to cook from the outside, as the food molecules on the outside absorb the microwaves and then conductively transfer the energy to the inside. Some plastics are not microwave-safe, and the hot food can heat the plastic to the melting point, or the heated plastic might leach chemicals into the food. For this reason, minimizing the use of plastics when microwaving food is a good idea. Metal or gold rims on mugs or plates are incompatible with the microwave as they can reflect the waves and cause sparks or fires.

Infrared (IR) radiation has a wavelength ranging from 0.1 to 1,000 microns (a micron is 0.000001, or 10^{-6} m). While we do not "see" IR radiation, we can detect its presence as objects warm in its presence. Your morning toast cooks by IR radiation from the internal heating element of your toaster.[5] Another IR cooking technique is broiling food in an oven, on a spit, or over a campfire, where IR radiation is absorbed, and the food browns. In recent years, IR grills have become popular. Advantages are high energy efficiencies from consistent heat across the grill and faster cooking times while keeping food moist. IR lamps in a restaurant also keep food warm until served to consumers.

Transfer of Heat by Induction

Induction cooking is related to radiative cooking but has unique origins and applications. In induction cooking, electricity passes through a conductive coil of wire that sits just below the ceramic or glass stovetop. The passage of electrons through the coil creates a magnetic field perpendicular to the coil and the surface of the stovetop. Electronics in the unit transform the current sent to the coil from the 60 Hz typical of house current to about 10,000 Hz, and the magnetic field created oscillates in synch with the electric field. Iron atoms in stainless steel and cast iron or at the core of some coated cookware are magnetic and responsive to this magnetic field.[6] As the field switches direction, the iron switches its magnetic orientation, creating friction as the atoms reorient 10,000 times per second. Temperatures rise because of this increased thermal energy. Copper, aluminum, and most ceramic cookware is not magnetic and is unsuitable for induction cooking. One of the advantages of induction cooking is its responsiveness: heating occurs rapidly, and when turned off, the burner cools immediately. The close contact between the burner and the cookware also results in less lost heat to the surroundings and fewer pollutants in the kitchen. Since nothing gets hot except the pot, it is unlikely to cause burns that can occur with electric or gas stoves.

Induction cooking may be the cooking method of the future. Recommendations for minimizing exposure to stray magnetic fields created from induction stoves include maintaining safe distances from the stovetop, using compatible cookware, and centering the cookware on the burner.[7] The slight risks of harm to children, pregnant women, and individuals with pacemakers or other implanted electric devices are minimized by following safety recommendations. Recently, there has been much discussion surrounding the safety of gas stoves and the advantages and disadvantages of gas stoves compared to induction or electric ones.[8] It is important to note that gas stoves do not vent to the outside, so pollutants such as nitrogen dioxide, carbon monoxide, and formaldehyde may be released into the home. Some of these pollutants can cause respiratory illnesses or exacerbate existing ones, and none are good for the environment. On the other hand, induction stoves are expensive and require cookware that contains magnetic material. While no action on gas stoves is imminent, federal regulatory agencies are considering a ban on installing them in new constructions—opting for the safer option of electric or induction.

Try It Yourself! Explore the effects of heat on food by following the instructions in **Experiment 1: Transforming Food with Heat** in culinary explorations in Part IV.

Cryo-Cooking—What Happens When We Remove Thermal Energy?

Grab a hot pot from the stove without a potholder, and you will feel pain as your skin turns red and blisters. While you might think frostbite is very different, it produces the same physical symptoms: redness, blistering, and pain. In both cases, extreme temperature damages proteins in the skin. When we use low temperatures to transform food, what happens to the molecular components of food at high temperatures bears similarities to what happens at low temperatures. The proteins designed to be folded and functional at body temperature fall apart and denature at both high and low temperatures. The solidification of most of the liquids and oils in foods at low temperatures makes for interesting textures. It affords manipulations such as turning maple syrup into maple taffy (sugar on snow) or using liquid nitrogen to turn olive oil into olive oil lollipops or fresh herbs into fine powders that provide additional flavor and surprising textures.

Cold-temperature cooking requires a heat sink that can absorb energy from food. Low-temperature materials such as liquid nitrogen (-196°C), dry ice (-80°C), and even snow (0°C) provide this energy sink. Materials such as Styrofoam can safely contain cryogens and keep them cold. Food containers must be strong enough not to crack at low temperatures. Stainless steel bowls, wooden spatulas, and metal spoons withstand low temperatures but can burn your skin when used without proper protection (goggles and gloves). Safety training should always accompany the use of cryogens. In many of today's restaurant kitchens, a container called a Dewar prevents the evaporation of liquid nitrogen and ensures the safety of chefs.

Try it yourself! If transforming foods using extremely cold materials sounds interesting, try **Experiment 2: Transforming Food by Removing Heat—Cryogenic Cooking**, in culinary explorations in Part IV.

SECTION 2: COVALENT BONDS AND NONCOVALENT INTERACTIONS

Cooking food often breaks bonds or other interactions holding the food molecules together. Sometimes cooking will create new bonds and interactions. Each type of bond or interaction has its origin and a typical range of strengths and lengths. Knowing this makes you a more skilled chef. When you understand what bonds need breaking, you can make a better choice of weapon—whether heat, cold, acid, salt, pressure, or microbes—and you can adjust the time, temperature, and maybe even the seasonings to get the desired results.

First, the strongest bonds in our foods are covalent bonds, which involve two atoms sharing at least one pair of electrons in their outermost or valence shells. Atoms themselves, you may remember, have a positively charged nucleus and are surrounded by negatively charged electrons. Atoms are neutral, which means the negative charge of the electrons balances the positive charge of the nucleus. The electrons can be thought of as being packed into different concentric shells. Only the outermost, or valence, electrons interact when two atoms form a bond. Each atom is attempting to achieve a filled valence shell, and when two atoms each share an electron, the pair of electrons gets each atom closer to achieving this stability. Atoms can share one pair of electrons (a single bond), two pairs of electrons (a double bond), or three pairs of electrons (a triple bond) to get this stability.

Bond energies represent the amount of energy necessary to break the bond. It is sometimes possible to use enough energy when heating food to break the bond. Typical single bond strengths are about 100 kcal/mole.[‡] Double or triple bonds will be even stronger. Bond length is the distance between the center of the two atoms. Covalent bond lengths are about 1×10^{-10} m (defined as 1 Angstrom, or 1 Å). The exact lengths will increase or decrease depending on the size of the atom and whether the bond is single, double, or triple. The more electrons involved in the bond, the stronger and shorter the bond.

The most readily broken covalent bonds in food are the very same ones we have previously described as the linkage bonds holding together the building blocks of our complex food molecules: glycosidic bonds for sugars, ester bonds for fats, and peptide bonds for amino acids. You may have already noticed that in each of these bond-forming reactions, the reactants lost a water molecule—a reaction known as dehydration. The simplest way to separate the complex molecule is to add a water molecule back again, a process known as hydration. Many cooking reactions involve this breakdown process, sometimes just by adding heat. If too much heat is needed to break the molecule apart, we might add an enzyme, a protein that acts as a catalyst that can speed up the reaction and require less heat. One example of an enzyme is invertase, which confectioners use to break down sugar when making candies with liquid centers. Another example of enzymes is the enzyme papain from the papaya fruit, which is used to break down proteins to tenderize meat.

Figure 3.4 lists other types of interactions categorized as noncovalent. These bonds can be longer than covalent bonds between the same two atoms. The strongest noncovalent interaction is an electrostatic attraction, or ionic bond, in foods like table salt (NaCl). Here the positively charged Na^+ ion strongly attracts the negatively charged Cl^- ion. Ionic bonds can be incredibly strong and almost impossible to break apart when the salt is a solid or gas, but put salt in water, and you know what happens! The electrostatic interactions break up easily because the water itself is also polar. A positive or negative ion can also

‡ A mole is a term used to describe a specific number of atoms or molecules, explained in greater detail in Chapter 4.

Name	**Example**	**Driving Force -** E (kcal/mole)	**Example**
Ion-Ion	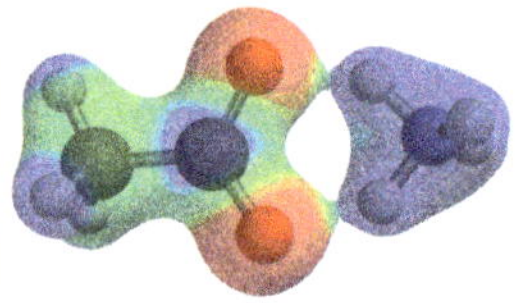	Ion charges E variable but large)	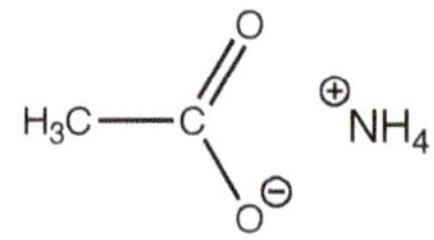
Ion-Dipole	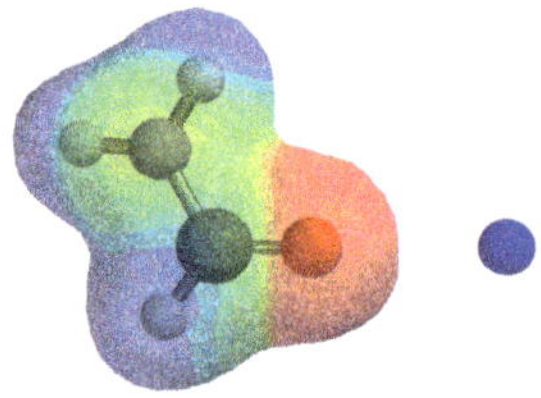	Charge + dipole E = 10-150	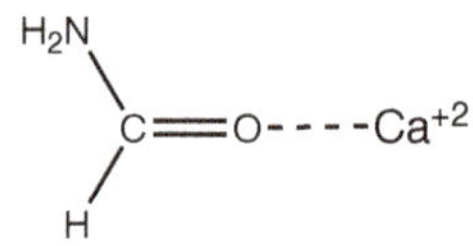
H-Bond	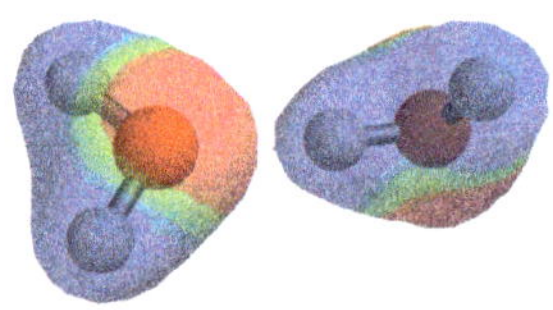	Large dipoles of H bonded to O, N or F E = 2-10	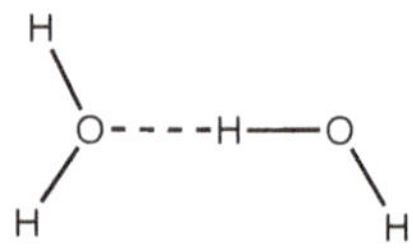
Dipole-Dipole	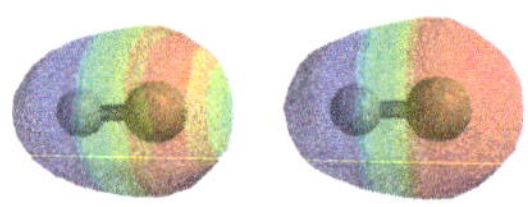	Dipole moments E = 1-6	
Dispersion Van der Waals		Polarizable surfaces E = 0.01-10	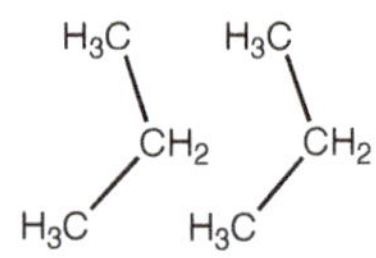

Figure 3.4 Noncovalent Bonding Interactions

interact with a neutral molecule with a polarized electron cloud, an interaction called ion-dipole. One of the more important noncovalent interactions is the hydrogen bond, or H-bond. An H-bond forms between hydrogen and another atom with a substantially greater electronegativity than hydrogen. Electronegativity measures how desperately an atom wants an electron and depends on how close it is to completing its outer shell of electrons. Atoms commonly found in food, such as nitrogen (N) and oxygen (O), have greater electronegativity than hydrogen (H), so they pull that electron toward them and make a H-bond, shown often as a dotted line N---H, or O---H. Hydrogen bonds have strengths of 2 to 10 kcal/mole and are very important in the structure of proteins, where they help stabilize the alpha helices in the three-dimensional structure.

SECTION 3: COOKING METHOD DEPENDS ON COMPOSITION

We know that most of the digestible composition of our food falls into three macronutrient categories: carbohydrates, fats, and proteins. Foods also contain water, salt, vitamins, minerals, and many small molecules known as micronutrients. Fiber is essential, even though it has no calories. Chefs train to understand how each of these components and the whole food responds when heated. Here we will examine this food transformation at both a chemical and an application level.

How Does Heat Transform Carbohydrates and Sugars?

As we saw in Chapter 1, sugars and carbohydrates are characterized by having a carbon, hydrogen, and oxygen ratio of one carbon to two hydrogens to one oxygen, or C-H_2O: "carbo-hydrate." The monosaccharides, simple sugar units with four, five, or six carbons, combine to form a disaccharide. The new bond created is called a glycosidic bond. It forms when an oxygen atom from an alcohol (-OH) on one sugar attacks a carbon atom on a second sugar. When the alcohol (-OH) on that second carbon picks up a hydrogen ion, or H^+, the newly formed dimer loses a water molecule, H_2O. Figure 3.5 shows sucrose, or table sugar, a disaccharide made of two sugar units (one glucose and one fructose) connected by one glycosidic bond. A polysaccharide contains many such units, each connected to the next by a glycosidic bond. Since the formation of each glycosidic bond results in the loss of a water molecule, the process is known as dehydration.

The process is reversed when disaccharides and polysaccharides break down; a water molecule adds to the bond and breaks the polymer into smaller units. This process is known as hydration or hydrolysis. In cooking with sugars, often the first step is the hydrolysis of that bond, which can be followed by

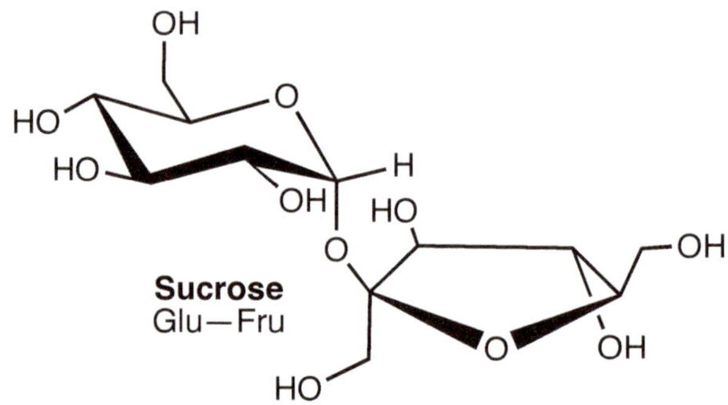

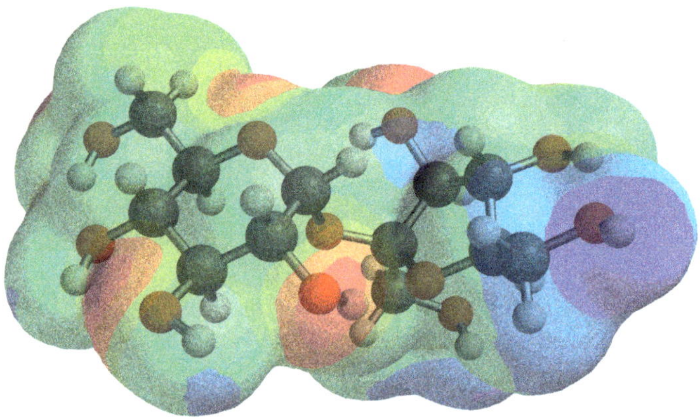

Figure 3.5 Table Sugar, or Sucrose, Is a Disaccharide Made up of Glucose and Fructose

The molecule is shown in both line drawings and as a surface charge rendered image. Note the rainbow of colors, with blue and red surfaces showing charge polarization alongside green surfaces showing neutral charge density. This molecule is soluble in polar solvents like water and alcohol but not in nonpolar solvents like mineral oil.

further bond breakage or rearrangement. This section considers some of the most common molecular transformations that occur when cooking with natural and artificial sweeteners.

Natural Sweeteners

In baking, chefs have an array of sweeteners to choose from. Each kind of sweetener has a different taste and different physical properties. Why are certain types of sugar used in particular recipes? While taste is a primary factor, the choice of sweetener is often determined by the colors and consistencies desired in the finished product. While honey, fruit, and maple syrup are great natural sweeteners, most bakers prefer the crystalline white sugar that is typically purified (refined) from sugar beets or sugarcane by extraction with alcohol. Some bakers, bartenders, and confectioners may need sugar that dissolves more quickly, so they opt for castor sugar or superfine sugar, in which the sugar crystals are finely ground. Confectioner's sugar is made of sugar crystals ground down to a powder to which cornstarch has been added to prevent clumping. This type of sugar works best for frostings or candies that contain little or no flour and where a smooth consistency is desirable. Brown sugar has residual molasses (or approximately 3.5% for light brown and 6.5% for dark brown) that coats the sugar molecules and produces an earthy caramel-like flavor. Molasses and treacle are essentially by-products of sugar refining. They contain residual sweetness but are unique for their deep, earthy flavors, important in producing candies such as toffee and licorice. Cornstarch, a sugar polymer, can be hydrolyzed and broken down into corn syrup (mostly maltose and glucose), a remarkably inexpensive sweetener. High fructose corn syrup, with its high shelf stability and low cost, is a rapidly growing source of commercial sweetener present in most processed foods.

The temperature at which sugar starts to break down depends on the chemical makeup of the sugar, as shown in Table 3–2. When water is present, it is not difficult to break down complex sugars into their monomeric states by adding a water molecule—a prerequisite to bond breakage to form simple sugars. But what happens when completely dry sugar is heated? Typically, sugars degrade in a process known as caramelization, which is discussed in greater detail later in this chapter. Further breakdown of the monosaccharide creates many different products that, unlike crystalline table sugar, are colored brown, some of which smell wonderful and taste even better. Voilà, caramel! One such breakdown molecule, diacetyl, has a butterscotch aroma and flavor, and another, hydroxymethyl furan (HMF), smells and tastes sweet. But beware! Too much heat, and the molecules can decompose to carbon, water, and other burnt-smelling, nasty-tasting by-products. Fructose is a simple monosaccharide in fruit and honey that begins to degrade at a much lower temperature than table sugar—105°C compared to 170°C—so foods baked with honey often turn brown.

In the confectionery world, when one might wish to create a liquid candy center, one can add the enzyme invertase to the sugar, and the hydrolysis of the sugar happens enzymatically at room temperature with no risk of the products further reacting to begin the tricky caramelization process. The process involves adding a minute quantity of the enzyme to a sugar paste made from table sugar, cornstarch, and water. The whole cherry is then embedded in a small patty of the sugar paste. Finally, the cherry-coated sugar patty gets dipped in chocolate. Over the next 24 hours, the enzyme transforms the sugar paste into a delectable liquid that oozes from the chocolate when one takes a bite of the candy.

Maltose, shown in Figure 3.6, is a natural disaccharide created when grains such as wheat or barley are softened and germinated in water by the action of

Table 3–2 **Different Sugar Caramelization Temperatures**

Sugar	Caramelization Temperature °C
Monosaccharides	
Fructose	105
Glucose	150
Galactose	160
Disaccharides	
Sucrose	170
Maltose	180
Isomalt*	190

SOURCE: Data from Harold McGee, *On Food and Cooking: The Science and Lore of the Kitchen*, Completely rev. and updated (New York: Scribner, 2004).
*Isomalt is not a sugar but a sugar derivative (see Figure 3.6).

the enzyme amylase on the storage polysaccharide amylose. When added to liquid desserts, this sugar provides a unique baked flavor and a sweetness that is a third of that of sucrose. Maltose breaks down at a high temperature of 180°C. Developed as a nutritional-based supplement for infants and invalids, it turned out people liked the flavor. Today we see it in malt beers, malted candies, and malted milkshakes.[9]

Storage polysaccharides such as glycogen in animals or amylose or starch in plants add caloric value to food and change the consistency and palatability of the cooked food product but do not add to the sweetness. Starch (amylose) is a solid at room temperature, but it can become soluble in hot water and leach out, leaving cloudy white scum on the pot when cooking foods such as rice or potatoes. Rice grains are often washed before cooking to extract soluble starch that might make the rice sticky or gummy. Starch is digestible and adds needed calories, so sometimes foods can be steamed, baked, or air-fried to capture a maximum amount of nutrition.

Artificial Sweeteners

Isomalt, shown in Figure 3.6, is an artificial sweetener used to create elaborate sugar sculptures. The first step in its synthesis from beet sugar is an isomerization involving moving electrons around the molecule, with some carbon atoms being oxidized and others reduced.[§] While Isomalt is safe to eat, it is not digested and does not affect insulin levels. It can cause some gastric distress as

§ One pneumonic to help you remember oxidation and reduction is "**LEO** the lion says **GER!**" for "**L**osing **E**lectrons **O**xidation" and "**G**aining **E**lectrons **R**eduction."

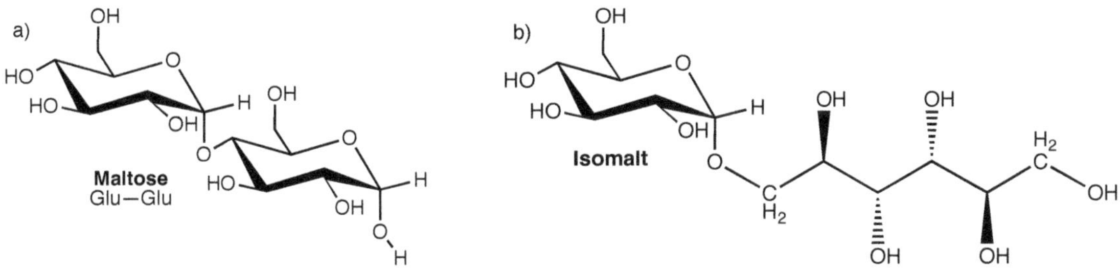

Figure 3.6 Two Sweeteners, the Natural Sugar Maltose and the Artificial Sweetener Isomalt

The skeletal structure for the disaccharide maltose, which is composed of two glucose molecules, is shown alongside the artificial sweetener Isomalt. A chemical rearrangement of atoms and electrons transforms the natural sugar maltose into the artificial sweetener Isomalt.

it passes right through you. Its chemical structure resists the breakdown associated with table sugar, and when it melts at 190°C, it can be fashioned into incredible shapes to make cake toppers and candies. These properties make it a favorite for decoration and novelty in professional and amateur cooking competitions.

As a result of sugar rationing, artificial sweeteners were widely developed during the past century's great wars. Saccharin (Sweet'n Low), sucralose (Splenda), and aspartame (Equal) all have zero calories and are not processed by our bodies. These are the sweeteners of choice for manufacturers of diet sodas and for consumers who must watch their sugar intake. The sweeteners do not, however, work well in a baking context because they lack the texture and granularity of sugar necessary for creating texture and color in the baked product. Studies show that hazardous polychlorinated compounds are formed when sucralose is heated in a water bath.[10]

In 2023, the World Health Organization (WHO) announced that it would list aspartame as possibly carcinogenic to humans. Other agencies were quick to challenge this classification. For example, the American Cancer Society wrote, "Scientific evidence has continued to support the Food and Drug Administration's (FDA's) conclusion that aspartame is safe for the general population when made under good manufacturing practices and used under the approved conditions."[11] "Safe conditions of use" translates to less than 40 mg/kg/day, or less than fourteen cans of 12-ounce soda daily. Most scientific agencies agree that more research needs to be done.[12]

A highly sought-after goal of food scientists is to develop a sugar substitute with fewer calories that has no unwanted health outcomes, is derived from natural products, has no aftertaste, and behaves like sugar in baking. Three new products show promise. The sugar substitute, allulose, is a natural grain sugar that tastes slightly less sweet than sugar (67%) but has only 10% of the calories.

Since most of the allulose consumed is not absorbed but excreted, it presents a compelling alternative sweetener as it provides the sweetness of sucrose but does not require the consumption of artificial ingredients. Incredo is another new sweetener trying to take its place in our kitchens and restaurants. Taking a different approach, the manufacturers of Incredo add silica granules to ordinary processed sugar. According to the FDA, the silica additives are perfectly safe, and the result is a sugar-silica mixture that persists for a longer time on the tongue, so less sugar needs to be added. Chefs working with this say Incredo does not have an aftertaste and that few or no changes to cooking procedures, other than the reduction of sweetener, are necessary. A final new candidate in the world of sweeteners is Supplant, a sugar reclaimed from leftover fiber from plant waste, such as corncobs, oat fibers, and wheat bran, which are first ground and then processed enzymatically to break the mixture down into a dry white powder. It is reported to bake and taste like sugar but has fewer calories. Because it contains fibrous materials from plant waste, it is metabolized slower in the body, which minimizes the sugar spikes and troughs that can be unhealthy.

How Does Heat Transform Fats?

A quick review of the information in Chapter 1: Fats and oils include a group of related food molecules characterized by a high composition of carbon (C) and hydrogen (H) and a low composition of oxygen (O). A fundamental building block for all fats is a fatty acid, a long hydrocarbon chain of just H and C atoms, ending with a carboxylic acid group (-COOH). Bonds between the C atoms in the chain can be single (C-C) or double (C=C). All singly bonded carbon chains are called saturated fatty acids (SFA), chains with one double bond are called monounsaturated (MUFA), and chains with multiple double bonds are called polyunsaturated fatty acids (PUFA). A fat, or triacylglycerol (TAG), molecule is made when a small three carbon glycerol molecule connects to three separate fatty acid chains through a bond known as an ester bond, where the -C-OH on the glycerol combines with the -COOH of the fatty acid to form the ester product (-COOC-) and a water molecule. Loss of water causes a dehydration reaction in the formation of the ester bond.

When energy is added to fat, it can melt if it is a solid or undergo hydrolysis, oxidation, or decomposition as bonds are broken. If a double bond is present, it can undergo isomerization. This section also discusses using oils to cook foods by transferring energy by convection when blackening, browning, deep-fat frying, pan frying, shallow frying, stir-frying, or sautéing. It is crucial to know the smoke point when choosing an oil. Selecting meats and fish with the proper fat content affects the selection of the cooking technique, the final flavor, and even the price you can expect to pay. We will see how such factors as beef marbling or the insulation necessary for a cold-water fish relate to cooking and flavor.

When Fats Melt

Using heat to melt solid fats dispersed throughout meat is known as rendering—such as when you cook bacon. The product, flavored by the liquid fat in which it has been cooked and the browned meat, is delicious and can be eaten as is (bacon or lardons) or incorporated into other dishes in gravies or quiches. Solid fats like lard and butter create flaky layers in baked goods such as croissants and pie crusts. Butter is often chilled first, cut into dispersed cubes, and not thoroughly mixed with the flour. Handling is kept to a minimum. When placed in the oven, the solid butter melts, leaving air pockets contributing to the highly desirable flakiness in croissants or pie crusts. In other dishes, you may need melted butter. Butter is an emulsion of water, fat, and milk protein. As you add energy, the solid fat melts, and the emulsion breaks, and you can observe the separate layers of fat, water, and protein. Continuing to heat the melted butter can denature the milk proteins and create browned butter—which can be delicious and is used in baking to add darker flavors to the richness of the butter fats. If the milk solids and water are removed, you form ghee, a staple fat in Southwest Asian cooking. If you add too much energy, the butter will burn—and bonds are broken in the fat to create smaller molecules with off-flavors and aromas as the fat breaks down.

When Fats Are Used as Cooking Oil

Fats that are liquids at room temperature are often used when frying and sautéing to conduct heat from a hot pan to raw food in an even way that produces good browning and adds flavor and richness to the dish. Choosing healthy fat with the right flavor profile and smoke point is essential to cooking. The smoke point is the temperature at which the heated oil starts to smoke—and since off-flavors and aromas are produced, it is best to stay below the smoke point. As seen in Figure 3.7, fats with a high percentage of SFA (blue bars) that are found in solid animal fats, such as butter and lard, have a low smoke point. Fats with a higher percentage of MUFA (green bars) and PUFA (yellow bars), such as those from fruits, vegetables, and nuts, are liquid oils at room temperature and have higher smoke points. While the length of the carbon chains and the degree of unsaturation (PUFA vs. MUFA) contribute to the smoke point, a more significant factor is the presence of free fatty acids (FFA) that are breakdown products of the TAG. Refined oils have been chemically processed to remove the FFA, so they have higher smoke points. Safflower and corn oils have high PUFA and smoke points above 200°C, so they are often used for deep fat frying where the high temperatures of the oil can produce crisp brown french fries or fried fish. These oils are also almost tasteless, since refining removes all the unique flavor and aroma molecules. Flaxseed oil has high PUFA but is unrefined, so the presence of impurities lowers its smoke point.

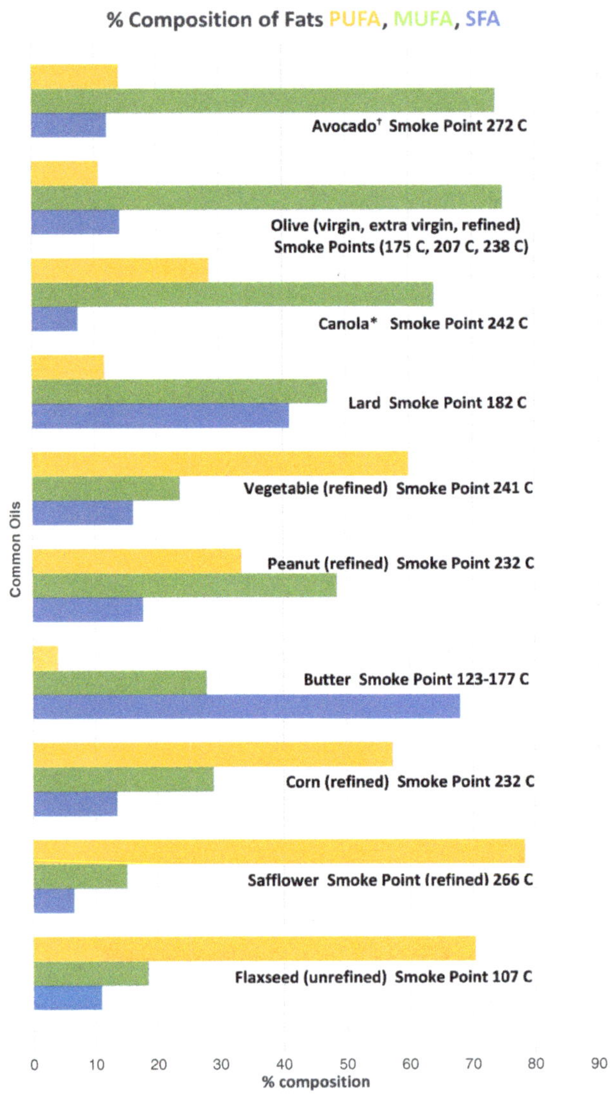

Figure 3.7 Comparison of the Composition and Smoke Points of Dietary Fats

Plant oils and animal fats contain different percentages of polyunsaturated fatty acids, PUFA, shown in yellow; monounsaturated fatty acids, MUFA, shown in green; and saturated fatty acids, SFA, shown in blue. The temperatures at which each of the fats begins to smoke, the smoke points, are listed next to the bar graphs. Fats with higher saturated fat concentrations (blue bar) tend to have lower smoke points. Oils and fats begin to break down at their smoke points, which can limit their usefulness when cooking at higher temperatures.

* Canola oil is a hybrid plant bred from rapeseed developed in 1970 in Canada

† Avocado oil has such a high smoke point because of its low FFA and high vitamin E concentrations.

Figure 3.7 shows that olive oil contains 74% MUFA, a C18 acid called oleic acid known to be a particularly heart-healthy oil. When harvested, olives destined to produce the high-value extra virgin olive oil (EVOO) must be processed quickly and cleanly. The fruit cannot have mold or bruises, or this will lead to off-flavors. Degradation of the olives starts as soon as they are picked, so speed is of the essence to minimize the breakdown of the ester bond that connects the FFA to the TAG. This FFA contamination leads to the lower smoke point of olive oil compared to other oils. There are strict rules for classifying an olive oil as EVOO; it must never have been chemically processed and must have an FFA of less than 0.8%. Oils with higher FFA are classified as the less expensive virgin olive and pomace oils. These oils have lower smoke points. An excellent EVOO can have a smoke point equal to some processed oils, yet retain the wonderful flavor, aroma, and health-giving compounds for which olive oil is famous. Alternatively, the olive oil can be chemically processed to remove these molecules, and the product, called Refined Olive Oil or Pure Olive Oil, will have a smoke point like other processed oils.

It is interesting to note that avocado oil has an almost identical fatty acid profile to olive oil. It is also a natural oil that does not need to be refined. Nevertheless, it has a smoke point of 271°C, much higher than EVOO and equal to or better than many processed oils. It turns out that the tough skin of the avocado and the fact that it can ripen in its tough shell after it is picked contribute to FFA levels close to zero in avocado oil and a correspondingly high smoke point.

When Bonds Break upon Heating

All bonds can be broken if enough heat is added; weaker bonds will break before stronger bonds. Fats and oils do not denature like proteins; they have fewer weak noncovalent bonds holding them together. The covalent ester linkage in TAG that holds the FFA to the glycerol can be easily broken down if water is present—which there almost always is. This reaction happens more quickly with heat, as the molecules have more kinetic energy at higher temperatures, and the extra jiggling around causes increased collisions between the molecules. The water added to the ester bond breaks off the free fatty acid, leaving a diacylglycerol (DAG) behind. A second water molecule can hydrolyze another ester bond, leaving another free fatty acid and a monoacylglycerol (MAG).

Another heat-facilitated reaction that can break apart TAG is that between molecular oxygen in the air and the hydrocarbon chain, especially if any double bonds react readily with oxygen and produce oxidation products that lead to chain breakage. The hydrocarbon chain begins to fragment at the smoke point, producing off-aromas and the smoke created by smaller hydrocarbon fragments.

When Bonds Isomerize with Heat

Fat derived from plants and animals contain double bonds (MUFA and PUFA) that all have the same geometry about the double bond, which is known as the "cis" conformation, as is shown by the monounsaturated fatty acid derived from oleic acid in Figure 3.8a. When in the cis orientation, the two hydrocarbon chains on either side of the double bond are on the same (cis) side of the double bond. By contrast, trans fats have a different geometry where the hydrocarbon chains are on opposite sides of the double bond. These two molecules are geometric isomers of one another—same molecules, different geometries. Both cis and trans isomers are shown in Figure 3.8a. High temperatures used in the refining of oils can lead to the double bond changing from its natural cis conformation to an unnatural trans conformation. This process is known as isomerization and leads to the production of trans fats, which are related to heart disease. Therefore, the FDA has banned the sale and use of oils containing trans fats.

Lycopene, a fatty acid–like molecule from tomatoes, is an antioxidant contributing to the tomato's status today as an exceptionally healthy vegetable.¶ Ironically, for nearly 200 years, this Central American fruit was called a "poison apple."[13] In the seventeenth and eighteenth centuries, people in Britain and her colonies were certain this food was deadly due to its classification as part of a family of plants that included deadly nightshade, a well-known poison. Over time, this uninformed taboo evaporated. Today the tomato is the fourth most common fresh vegetable and the most common canned vegetable consumed in the United States.[14] Lycopene is part of that healthy molecular profile. It has a long hydrocarbon chain but lacks the terminal carboxylic acid. The structure of lycopene, shown in Figure 3.8b, has many double bonds, but unlike natural fatty acids, the double bonds are all in the trans configuration. When heated, the double bond can isomerize to a cis conformation about one or more of its double bonds. Fortunately for us, cis lycopene is more easily absorbed in our gut. So you can feel good about eating another serving of spaghetti sauce, knowing that the lycopene has been isomerized just for you!

Intrinsic Fat Content of Food Is Transformed during Cooking

Wagyu beef, roasted goose, and roasted pig are succulent foods that make your mouth water. What do they all have in common? Fat. In each case, fat adds to the mouth-watering juiciness of the meat and adds value to the product. These foods provide a perfect balance of tasty meat and fat that bastes the food as it cooks, tenderizing it and providing browning reactions that

¶ Though botanically classified as a fruit, nutritionists classify the tomato as a vegetable.

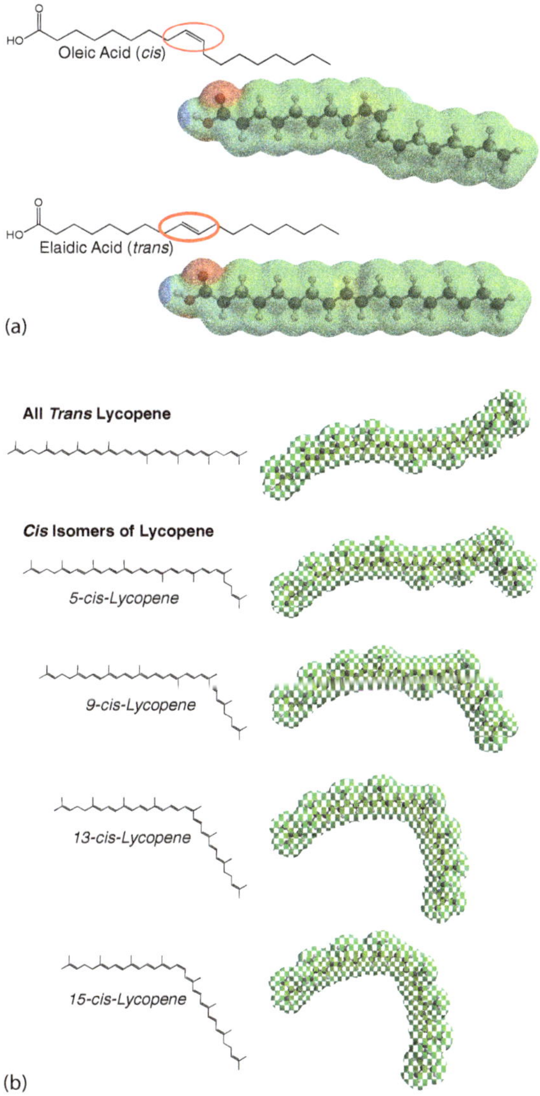

Figure 3.8 Isomerization of Double Bonds by Heat

a. The cis double bond of the oleic acid in olive oil can be isomerized to a trans double bond of elaidic acid as excess heat is added. The double bond is circled in red in the line drawings of the two molecules, showing where this isomerization occurs. The surface images highlight the change in three-dimensional geometry that results from the isomerization. Any appreciable levels of trans isomer in olive oil is evidence of the aging of the oil and poor handling of the product and would cause an extra virgin olive oil to lose this designation.

b. Lycopene in tomatoes exists as the all-trans isomer. Heating tomatoes (especially in olive oil) causes double bonds to isomerize, creating a host of different product molecules, a sampling of which are shown here (5, 9, 13, 15 cis). One beneficial result of this isomerization is that the isomers are better absorbed in the gut than the all-trans isomer, making your tomatoes even healthier.

add so much flavor. Consumers are willing to pay a lot for these foods: imported Wagyu olive beef, from beef cattle fed olives to provide even more delicious fat, costs as much as $400 per pound.[15] A certain degree of fat marbling in steak is a hallmark of the highest grade of meat. The slow roasting of a pig or a lamb, which can be particularly fatty, can produce a delicious dish as the fat is slowly rendered from the meat and naturally bastes it as it cooks. Certain pheasants, such as duck or goose, also have a large amount of fat associated with the richness of these meats.

Fish is a source of protein and is lower in fat than most meats, but not all fish are created equal. Fat content varies significantly from species to species: certain cold-water fish such as sardines, mackerel, blue fish, salmon, and some tuna are more oily than others. In cold-water fish such as sardines and mackerel, the oil provides insulation. Flakiness varies from one fish to another. The shorter muscle fibers of fish compared to land animals mean less collagen—3% (but variable) in fish compared to 15% for us terrestrial animals. When cooked, the collagen melts, leaving behind the flaky flesh. When fish such as salmon are overcooked or cooked too quickly, protein leaks out of the layers as the muscle fibers contract too quickly. This protein, albumin, can solidify into a white gooey substance when exposed to heat—unattractive but perfectly safe and healthy to eat. Cooking more slowly with the skin side down can help minimize this leakage. Tuna, by contrast, is a warm-blooded fish, and in some types of tuna, the fat can be seen marbled through the flesh—not layered as in salmon. When uncooked, such as in fatty tuna used to make sashimi, the fat provides a rich, creamy mouthfeel. The lower abdominal meat on the prized bluefin tuna—known as otoro tuna—can cost $200 per pound.[16]

How Does Heat Transform Proteins?

In Chapter 1, we learned that proteins are made of amino acids connected by strong covalent peptide bonds and that twenty different amino acids are used by our bodies to build proteins. These different amino acids have a common -N-Cα-C- backbone but different side groups. If the side group of the amino acid is small, charged, or polar, it will tend to be water soluble, and amino acids of this type are described as hydrophilic, or water-loving. If the side group of the amino acid is large and bulky or nonpolar, it tends to try to avoid water and is described as hydrophobic. In its natural, raw, uncooked state, a protein folds to bury the hydrophobic amino acids on the interior and leave its hydrophilic amino acids exposed to water. Each protein folds spontaneously into a unique three-dimensional shape depending on the sequence of amino acids. Hundreds of weak noncovalent interactions, listed in Figure 3.4, hold the shape together. We saw examples of the different shapes of proteins formed from four of the signaling hormones in Chapter 1. It is estimated that our bodies make 100,000 different kinds of proteins.

Since each protein has a unique sequence of amino acids, it will also have a unique folded structure with remarkable stability. The cooking process for most proteins disrupts those weak interactions, and the natural three-dimensional folding of proteins is disturbed. The added heat breaks the weak noncovalent bonds, unfolding the protein to a shape known as a random coil. We say that the protein has denatured. Random coils, like strings of cooked spaghetti, tangle up with one another, and the denatured protein is said to aggregate. While some chemical changes can make this process faster, it is essentially a physical change of the protein from a folded state to an unfolded state—almost like melting. Each protein has a unique unfolding temperature: some proteins are particularly unstable and do not need any heat beyond room temperature to unfold; other proteins can be heated to boiling at 100°C and still not unfold. The amino acid sequence is what determines the unfolding temperature.

While not important for consideration of cooking per se, a denatured protein is farther along on its digestive pathway as the protein's nutrition lies in the amino acids themselves. To access those amino acids, it is necessary first to unfold the protein so that enzymes can then access the stronger covalent peptide bonds holding them together. Once food enters the mouth, it meets up with the digestive enzymes in saliva, the acid in the stomach, and the other enzymes in the gut so that a protein is eventually broken down into component amino acids. Those amino acids will be used to make new proteins, burned for fuel, or repurposed in many other ways.

A Perfect Eggs-ample

Cooking an egg allows for a perfect and familiar example of how heat affects the structure of different proteins. Nothing could be simpler than boiling an egg. Evidence shows humans were eating eggs 10,000 years ago in Southeast Asia.[17] Today eggs are considered a global food. Lisa Bramen took on the challenge of describing one egg dish from each of eighty countries in her article, "Around the World in 80 Eggs!," published in *Smithsonian Magazine*.[18] Despite the egg's ancient and humble origins, the preparation of egg dishes occupies an essential place in culinary education: 100 pleats in a chef's white toque are said to represent mastery of the 100 different cooking techniques.[19] What can this ubiquitous food teach us about transforming proteins with heat?

The nutritional label for an egg shown in Figure 3.9 provides the first piece of the puzzle. One large egg weighing approximately 53 grams provides about 70 calories of energy, roughly 6 grams of protein, and 5 grams of fat.[20] A few of the other vital nutrients present in eggs are vitamin B12, or choline, necessary

Nutrition Facts	
Serving size	**1 egg (53g)**
Amount Per Serving	
Calories	**70**
	% Daily Value*
Total Fat 5g	**6%**
Saturated Fat 1.5g	**8%**
Trans Fat 0g	
Cholesterol 195mg	**65%**
Sodium 65mg	**3%**
Total Carbohydrate 1g	**0%**
Dietary Fiber 0g	**0%**
Total Sugars 0g	
Includes 0g Added Sugars	**0%**
Sugar Alcohol 0g	
Protein 6g	**12%**
Vitamin D 3mcg	15%
Calcium 26mg	2%
Iron 1.08mg	6%
Potassium 0mg	0%
Vitamin A	10%
Vitamin C	0%
Vitamin E	15%
Riboflavin	15%
Niacin	8%
Folate	15%
Vitamin B12	50%

*The % Daily Value (DV) tells you how much a nutrient in a serving of food contributes to a daily diet. 2,000 calories a day is used for general nutrition advice.

Figure 3.9 Nutritional Label for an Egg

Label for one egg weighing 53 grams shows the distribution of macronutrients (Carbohydrates, 1 gram; Fat, 5 grams; Protein, 6 grams) and micronutrient vitamins and minerals.

for fetal brain health and development; and vitamin A, vitamin D, thiamine, riboflavin, iron, phosphorus, and calcium—all of which are necessary for the growing embryo. The molecular structures of some of these nutrients are shown in Figure 3.10. Brown and white eggs have no intrinsic difference in flavor. The yellow color from lutein and the orange color from zeaxanthins and chicken food additives such as marigold flowers can make the yolks orange. Figure 3.11 shows the anatomy of an egg. A relatively rigid but semiporous shell

Molecule - Function	**Structure**
Vitamin B – choline Important in brain health & development	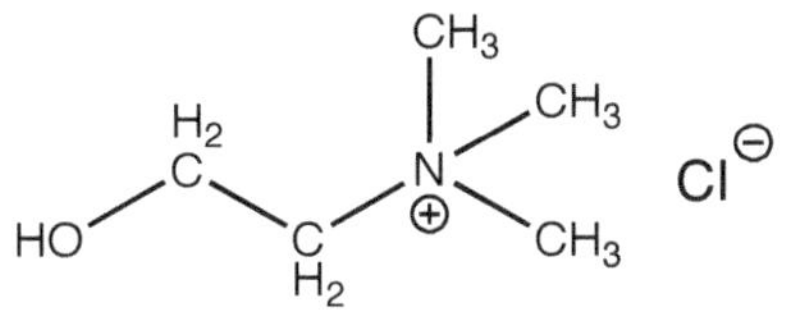
Zeaxanthin - Antioxidant that contributes to the orange color of yolk	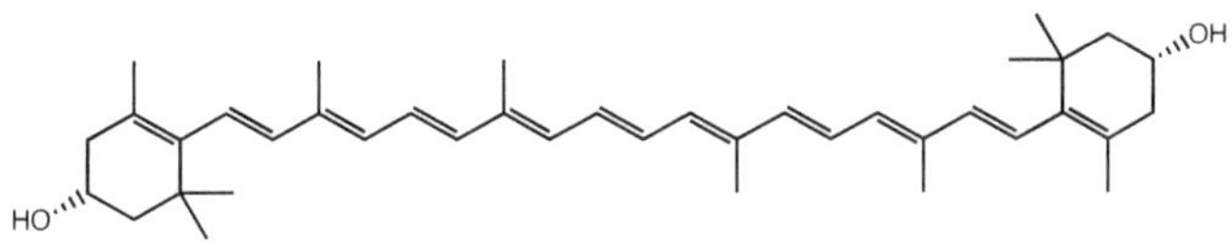
Lutein - Antioxidant that contributes to the yellow color of yolk	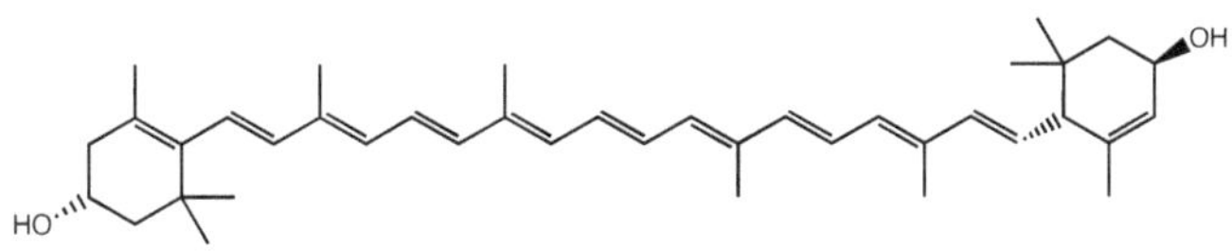
Cholesterol - Important in the structure of cell membranes	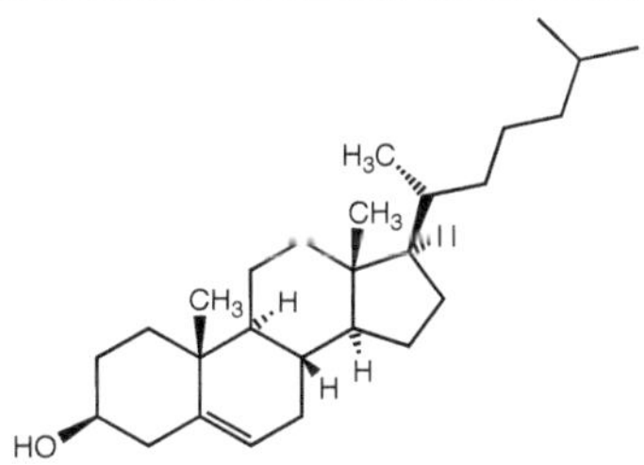

Figure 3.10 Important Micronutrients in an Egg

The names, functions, and skeletal structures for four (vitamin B, zeaxanthin, lutein, and cholesterol) of the hundreds of micronutrients found in eggs are shown.

protects the fragile interior of any avian egg. The shell contains calcium carbonate, the same chemical that makes up chalk and limestone. The interior has two primary compartments, the white and the yolk, surrounded by a double membrane and tethered to these membranes by the feathery threads known as chalazae. Ten percent of the mass of the egg white and 20% of the mass of the egg yolk is protein. We will consider the 6 grams of protein in the egg here.

Upon heating, there is a profound change in the structure of egg proteins. As shown in cartoon form in Figure 3.12, heat causes proteins to unfold (denature), stick together (aggregate), and form a randomly latticed gel. As a result, the egg transitions from a thick liquid to a semisolid that you can pick up with a fork. Creating an egg with a runny yolk (proteins still folded) while the egg white is solid (proteins unfolded and aggregated) is especially challenging,

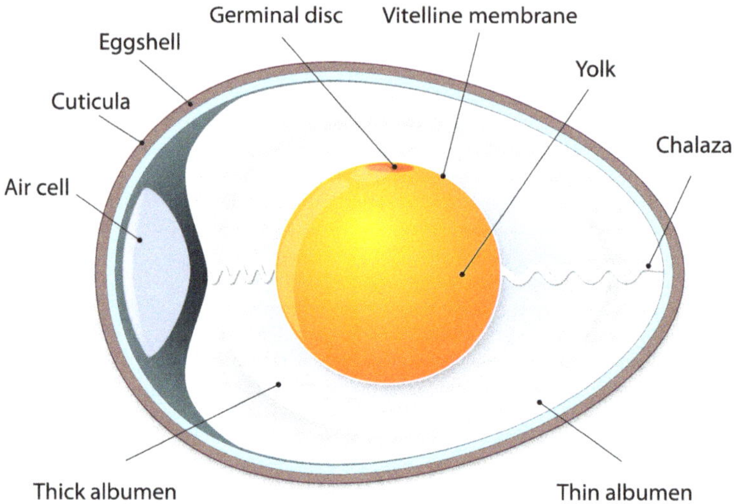

Figure 3.11 Anatomy of an Egg

The various parts of a chicken egg are shown (air cell, cuticula, eggshell, germinal disc, vitelline membrane, yolk, chalaza, and thick and thin albumen.) The albumen egg white is 90% water and 10% protein with a pH of 8.8. The egg yolk has a lower pH of 6.5 and less water (50%) but twice as much protein (20%) and 30% fat.

given the different denaturation temperatures of the various protein components. Cholesterol in the egg yolk helps prevent yolk proteins from aggregating. The aggregated proteins in a cooked egg form the matrix that holds water, which turns to steam when cooking and helps make cooked egg fluffy. When the egg is overcooked, the water has been steamed out of the protein matrix, making the egg rubbery. Salt and fat help hold the matrix together and make eggs fluffy by preventing the tight packing of the aggregated proteins. Since the composition of the yolk and white differ, the temperature at which each transforms from liquid to solid varies. Typically, egg whites become solid at a lower temperature than egg yolks.

Different denaturation temperatures partly result from different amino acid compositions and protein stabilities. In an egg white, the proteins are primarily ones that will protect the growing embryo, such as albumen, globulin, transferrin, and avidin. Albumen and avidin act as preservatives to help protect the growing embryo by binding to other foreign proteins. Transferrin is an iron scavenger that also helps prevent microbial growth. The egg yolk is the part of the egg that would contain the growing embryo if it were fertilized. The yolk has twice as much protein per g than the egg white. Since yolks also have the majority of the fat and cholesterol in an egg, it will not surprise you that they contain a high percentage of proteins that bind fats and cholesterol: low-

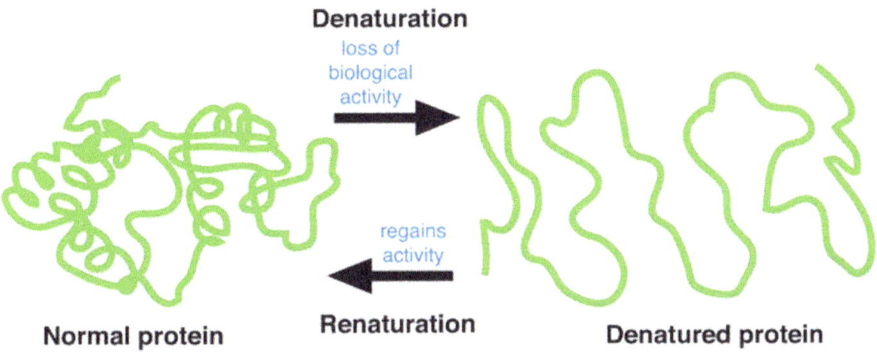

Figure 3.12 Denaturation of Egg Proteins

The proteins in eggs will unfold from their well-defined three-dimensional shapes when exposed to heat, acid, base, or salt. The unfolded protein strands can tangle, causing a clear solution to become opaque or even solid.

density lipoprotein, or LDL (65%), and high-density lipoprotein, or HDL.[21] Other major proteins are phosvitin, a highly phosphorylated protein essential in bone growth and in binding to calcium, and livetin, a water-soluble fraction that includes a mixture of immunoglobulin Y, albumin, and a glycoprotein.[22] Lipovitellins and vitellogenins are also present. They provide protein and lipid-rich nutrients for developing embryos and protect the growing embryo through their ability to recognize and kill bacteria and viruses.[23] These proteins and cholesterol cause yolks to become solid at higher temperatures.

The physical appearance of a boiled egg at various temperatures, shown in Figure 3.13, reveals how the proteins respond as the temperature is raised degree by degree. At each temperature point, the egg can equilibrate by waiting five minutes or more. At 57°C, an egg will remain raw with all proteins still folded. Notice that the white is still transparent, and the white and yolk are still liquid. Raising the temperature and allowing the egg to equilibrate at 60°C causes the white to set while the yolk remains runny. At 62°C, just 2° higher in temperature, you have prepared what many call the perfect egg to put on toast, with a soft white and a viscous but still liquid yolk. Add one more degree of heat, and the yolk turns creamy. At 75°C, your egg is hard-boiled with a rubbery consistency and with no liquid anywhere—perfect if you wish to prepare deviled eggs.

Another Example of Protein Denaturation and Aggregation—Shrimp

Another tangible demonstration of protein denaturation can be seen when cooking shrimp. As the translucent raw shrimp is heated, the first noticeable change is the color of the shell: the light blue-gray fades and is replaced by a

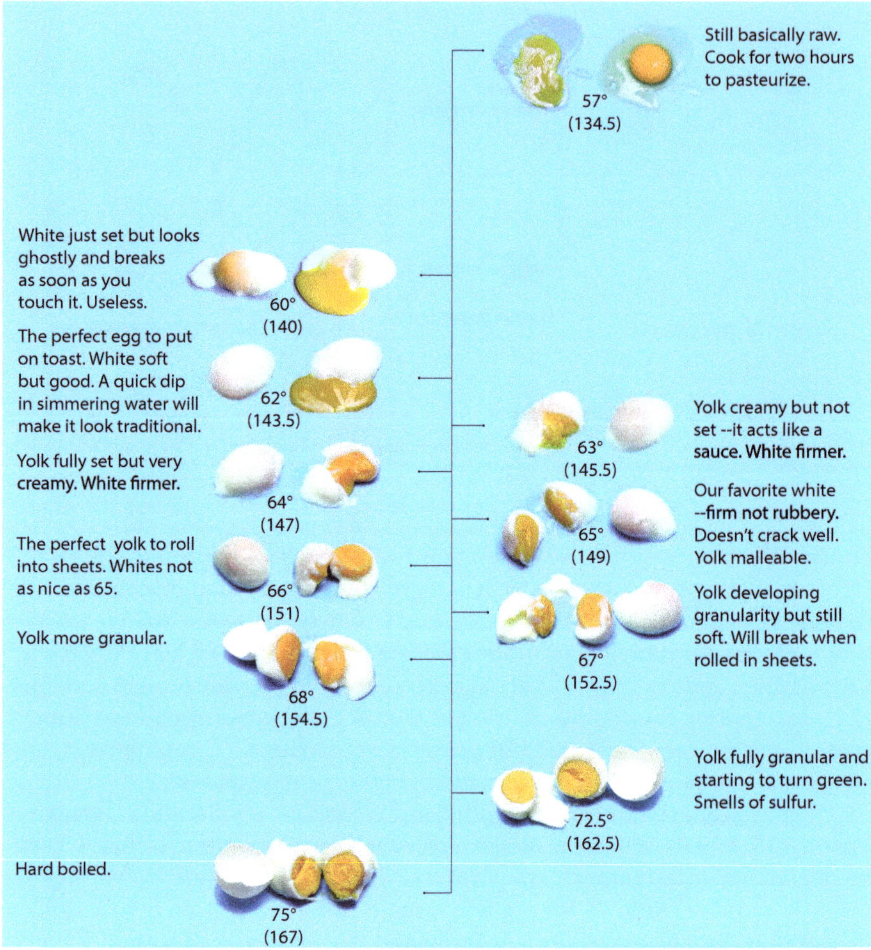

Figure 3.13 Eggs Cooked at Various Temperatures

The temperature at which an egg is cooked will result in different outcomes for the white and the yolk. At temperatures of ~60°C and higher, the proteins in the egg white will unfold, as evidenced by the egg white becoming solid and losing its transparency. Higher temperatures of ~64°C or more are necessary to unfold the proteins in the yolk and for the yolk to set—or become solid.

beautiful apricot pink. In the shell of raw shrimp, a natural pigment that acts as an antioxidant called astaxanthin is bound to a protein.[24] As the shrimp is heated, the protein unfolds, releasing the astaxanthin. In solution, the pigment has a different shape and absorbs light of a different wavelength, making it look pink. In the flesh of the shrimp, another transformation is taking place. Here, just like in the egg, the folded proteins absorb heat, changing from translucent (uncooked) to milky white (perfectly cooked) to solid white (overcooked). Muscle fibers in the shrimp also change shape as heat is added—in this case, shrinking and causing the shrimp to adopt greater and greater curvature from slight (uncooked) to crescent (perfectly cooked) to circular (overcooked). So by paying careful attention to the shell color, the flesh opacity, and the curvature of the crustacean, you will be rewarded with perfectly cooked shrimp, not rubbery or tough, with a beautiful pink shell.[25] A similar denaturing process can be induced by adding acid to raw shrimp, which is at the heart of a type of food preparation known as ceviché, which is discussed in the next chapter.

SECTION 4: MORE COMPLEX CONSIDERATIONS IN COOKING

Several types of processes can occur in foods that lead them to turn brown. These browning reactions are of two types: enzymatic browning, which occurs when fruits and vegetables are sliced open; and nonenzymatic browning, such as the caramelization reactions we saw with heated sugars and a second reaction known as the Maillard reaction.

Enzymatic Browning

While not caused by heating, enzymatic browning reactions are an essential starting point for a discussion of the browning reactions in general because they are familiar and have similarities with the food transformations caused by heat. Enzymatic browning reactions involve the oxidation of the amino acid tyrosine, which is catalyzed by the enzyme phenolase. In whole unbruised fruit, the enzyme is found inside the chloroplast of a plant cell where the oxygen concentration is very low. When fruit is cut, the cell's contents are released, exposing the enzyme to air—which is about 20% oxygen. Figure 3.14 shows the molecular transformations that take place as tyrosine is oxidized. First, oxygen is added to the six-membered ring of tyrosine, making hydroxy tyrosine, also called dihydroxyphenylalanine, or DOPA. In subsequent reactions, more electrons are lost to the oxygen, creating double bonds between carbon and oxygen. The molecule cyclizes, loses a CO_2 molecule, and becomes the indole quinone shown. These small subunits then polymerize to form melanin, a larger structure with alternating single and double bonds, which absorbs light and looks brown.[26]

Enzymatic browning can be minimized in a few ways. First, adding lemon juice or other acids, such as citric acid, to the surface of the cut flesh will slow down or denature the phenolase enzyme. Second, it is possible to microwave the product—just long enough to denature the phenolase enzyme but not long enough to "cook" the fruit or vegetable. Finally, it is possible to carefully cover the exposed fruit or vegetable with plastic wrap to minimize exposure to air or,

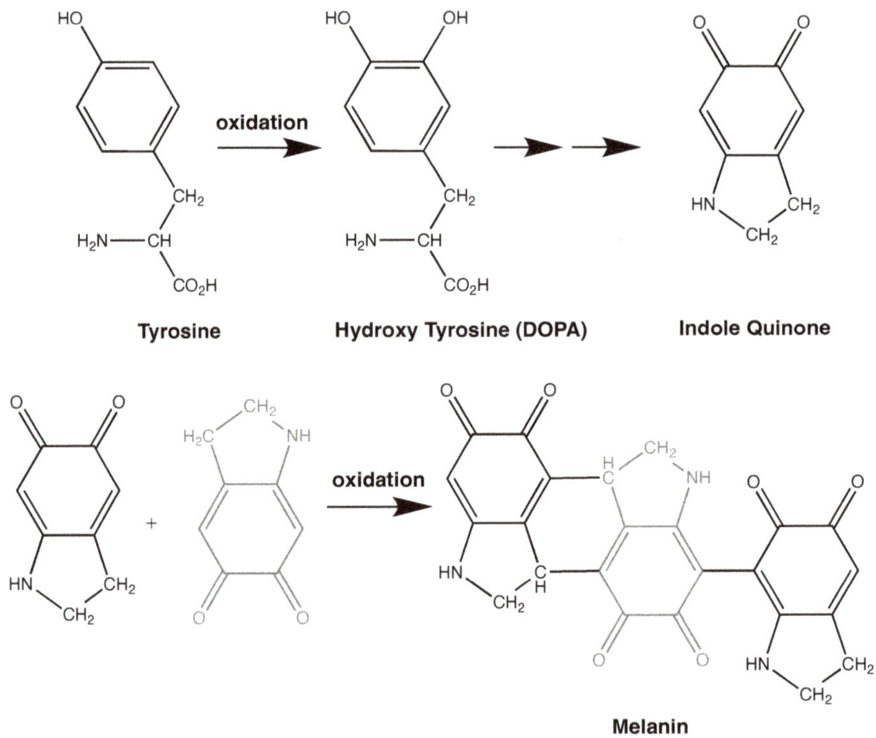

Figure 3.14 Enzymatic Browning Reactions

Sequential chemical reactions that transform the amino acid tyrosine into the pigment melanin through a series of oxidations and rearrangements. Melanin absorbs light across the visible spectrum, so materials with melanin appear black or brown. The enzyme phenolase catalyzes the first oxidation step.

if in cooking, to prepare it by the sous vide method. Some hybrids or genetically modified (GM) apples, such as the Arctic Granny and the Arctic Golden, have very low levels of the phenolase enzyme and are famously resistant to browning.

Nonenzymatic Browning Type I: Carmelization

Earlier in this chapter, we learned that the temperature at which a sugar begins to decompose depends on the structure of the sugar and that this is helpful information for baking and making confectionary creations. Here we will consider this degradation reaction in more detail. The reactions begin with hydrating the linking glycosidic bond in a disaccharide when heat is added, as shown in Figure 3.15. As can be seen, the disaccharide sucrose is first broken down into its components: sugars, glucose, and fructose.[27] Further heating causes glucose to lose three water molecules and be converted to maltol, which has a toasty odor. Heating also causes fructose to lose water molecules and be converted to hydroxymethyl furfural (HMF), which is responsible for the signature odor of caramel. Other breakdown products, such as diacetyl with an intense buttery flavor, add to the aroma. These reactions are extremely temperature sensitive—as anyone who has ever made caramel knows—one second too long—two degrees too high—and you have burnt sugar. Fortunately, some people love the flavor of burnt sugar, as evidenced by the many recipes for burnt sugar frosting or cookies. Suppose you do not have the patience to wait for your onions to caramelize. In that case, these reactions can be accelerated by adding an acid such as lemon juice or a base such as sodium bicarbonate.

The brown color of heated sugars is poorly understood, and while we know some details about the composition of the compounds that are formed, we do not yet know the structures. The three classes of polymers formed are caramelans $C_{24}H_{36}O_{18}$, $C_{36}H_{50}O_{25}$, and $C_{125}H_{188}O_{80}$, each with a darker and darker color as weight loss due to dehydration and other factors increases from 9% to 15% and 22%.[28]

Nonenzymatic Browning Type II: Maillard Reaction

One of the most frequently named reactions in all cooking is the Maillard reaction. Louis-Camille Maillard was a French researcher and physician who first observed and wrote about the reaction between sugars and amino acids in the presence of fat in 1912 in Paris. In 1953, John E. Hodge, an African American chemist working in Peoria, Illinois, established the mechanism that allowed scientists to understand how the reaction works.[29] In an article written in 2012 to celebrate the 100th anniversary of the Maillard reaction, the science writer Sarah Everts reported that Vincenzo Fogliano, a food chemist at the University of Naples Federico II, said, “Maillard discovered the reaction, but Hodge understood it.” In fact, because citations of Hodge’s paper far outnumber those of Maillard’s, there has been some discussion of renaming it the Maillard-Hodge reaction, Fogliano said. [30]

The Maillard reaction shown in Figure 3.16 occurs between a sugar and an amino acid at temperatures between 110°C and 150°C. The nitrogen in the amino acid ($-NH_2$) reacts with carbon in the sugar that is doubly bonded to oxygen ($-C{=}O$), causing the release of water from the sugar. The addition compound is unstable and rearranges to form a ketose amine ($-C{=}N$) known as the Amadori compound. This intermediate can go on to produce many different products depending on conditions such as pH and temperature, as well as the identity of the amino acid that reacts with the sugar.[31]

Molecules that produce flavors and odors are derived from these breakdown products. A few examples are shown in Figure 3.16. Some give rise to the toasty

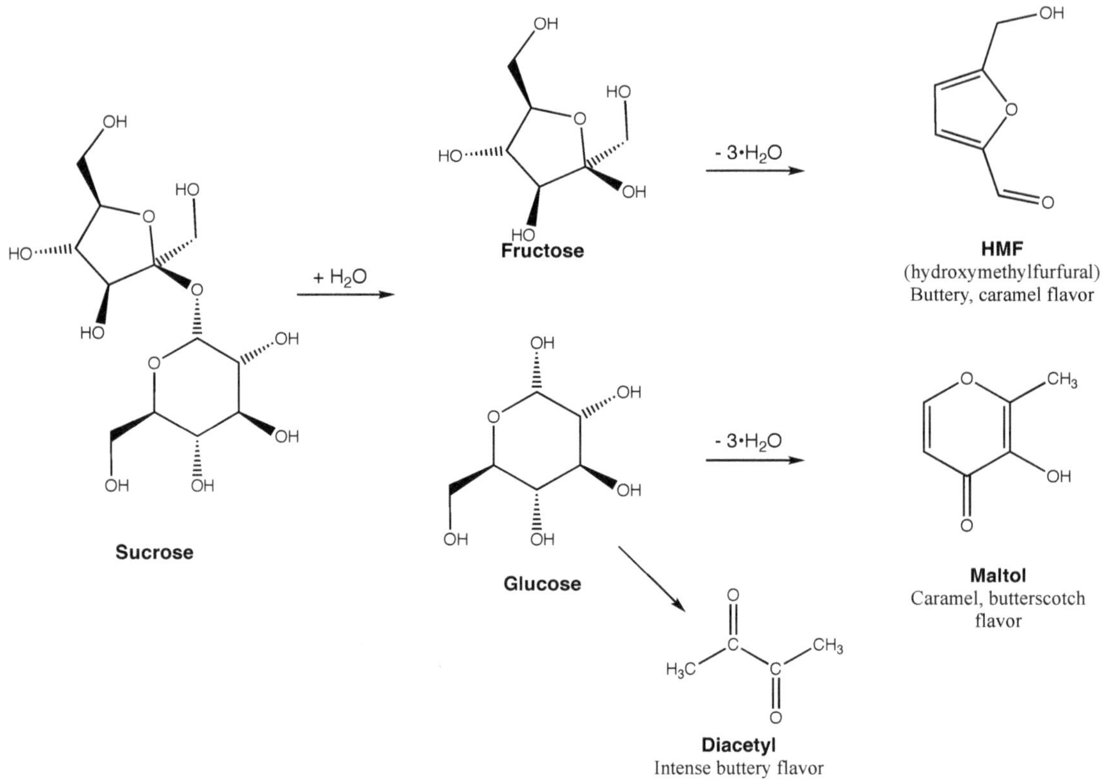

Figure 3.15 Caramelization Reactions

The glycosidic bond in disaccharides like sucrose can be broken by the addition of a water molecule, and with heat, the individual monosaccharides break down into a host of aromatic and flavorful smaller molecules such as HMF that give rise to the buttery caramel flavor.

sweet aroma from furans and the meaty smell from thiophenes. One such furan (with one N atom and four C atoms in the five-membered ring) is already familiar to us: HMF, which is also created in the caramelization reaction. Thiophenes (with one S atom and four C atoms in the five-membered ring) are formed when the reacting amino acid, such as cysteine and methionine, contains a sulfur atom. Six-membered rings are also created that give rise to aromas and flavors of cereal and grain (acyl pyridines with one N atom and five C atoms in the ring) and toasted flavors (pyrazines with two N atoms and four C atoms in the ring). Hundreds of compounds like these are formed, and the complex mixtures of these volatile compounds enter our noses and elicit the pleasant anticipation of having a lovely piece of fresh bread or grilled meat.

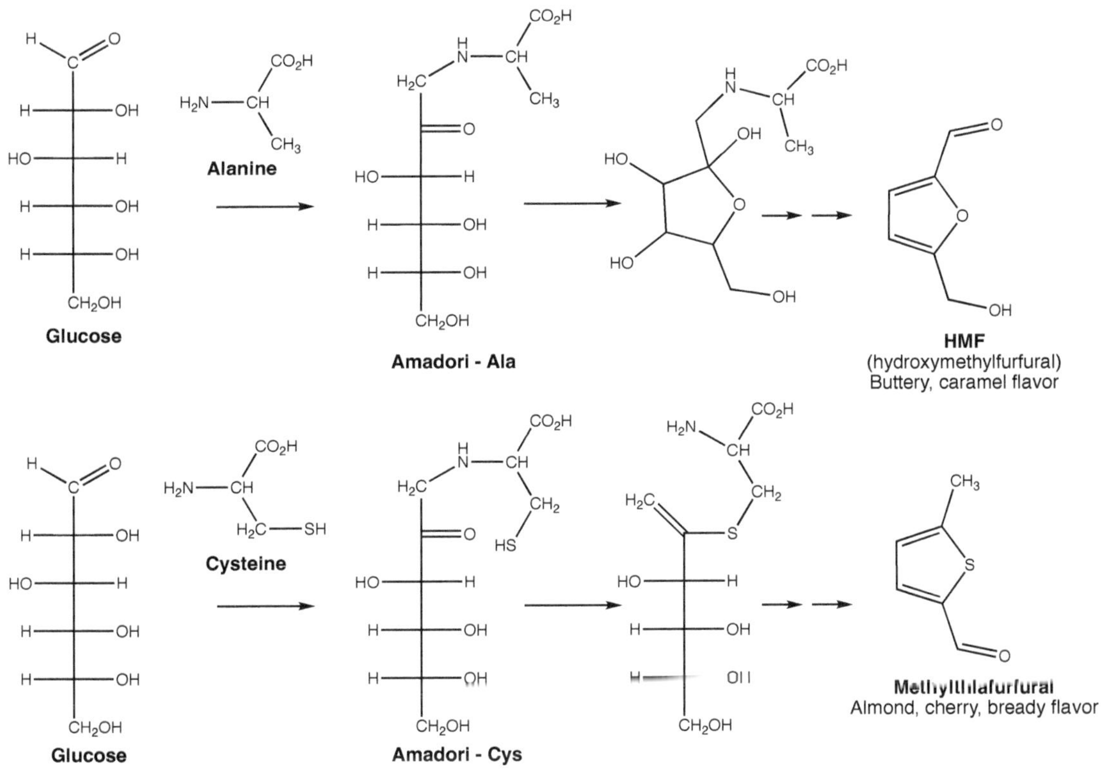

Figure 3.16 Maillard Reactions

Sugars and amino acids can react when heated to produce molecules with various delicious flavors and wonderful aromas. Two of the many examples are shown here: glucose plus an alanine amino acid can make the hydroxymethylfurfural molecule that has a buttery, caramel flavor, and the same sugar with a cysteine amino acid can produce the molecule methylthiafurfural that has an almond and bready flavor.

What gives rise to the color? Whether the golden crusts of bread, the grill marks on a steak, the deep rich brown of coffee, or even the amber color of our beer, all these brown colors that appear in foods that have undergone the Maillard reaction come from complex polymerization and complexation reactions. One such complex, between two Amadori compounds, is shown in Figure 3.16; as we have seen with the other browning reactions, the color of the complex results from the absorption of visible light made possible by the alternating double and single bonds spread out over so many atoms. Since the high temperatures necessary typically occur on the surface, brown crusts of bread and roasted skins on a turkey are surface features. In liquids such as coffee or beer, the color is dispersed throughout.

SUMMARY OF CHAPTER 3

We saw in this chapter how each type of macronutrient can be transformed with heat—the most traditional cooking method. We first identified the primary ways heat flows to a material: conduction, convection, and radiation. We learned that cookware is made from different elemental materials that conduct and store heat in different ways and that it is essential to choose the right cookware to achieve the desired results in the kitchen. Then we saw how heat transfer can change the physical state of food (from solid to liquid, for example). We learned that heat can also affect ingredients more complexly as the chemical changes involve breaking both weak intermolecular interactions and stronger covalent bonds. We learned that removing heat in cryogenic cooking can also cook foods. By adding or removing heat, we can unfold complex molecules, such as the proteins in a raw egg, that are stabilized by weak intermolecular forces. We can also decompose molecules such as sugar in caramelization reactions. We saw how oxidative processes can create delicious flavors in Maillard reactions. Oxidation can also degrade food, and we explored how those processes can be slowed down or stopped.

In Chapter 4, we will see how food can also be transformed with acid in foods such as ceviche, with microbes in fermentation, and with pressure.

Check out the two lab experiments, **Experiment 1: Transforming Food with Heat** and **Experiment 2: Transforming Food by Removing Heat—Cryogenic Cuisine**, in culinary explorations in Part IV.

CHAPTER

4

Transforming Food Using pH, Pressure, and Microbes

In the previous chapter, we saw how heating or cooling can transform food by changing chemical bonds. In Chapter 4, we will learn about additional methods to transform food using pH, changes in pressure (with heat), and microbial processing or fermentation. We will start in **Section 1** by considering the familiar concept of the acidity of food, and while the term pH might be somewhat less familiar, the two are related. Foods with high acidity have a low pH. Conversely, foods with low acidity are called basic and have a high pH. Most foods contain a balance of acid and base and are considered neutral, with a pH of around 7. We will learn how altering this balance can transform ingredients, producing more digestible and preserved foods without adding heat. To fully understand what is happening with acids and bases at a molecular level, we will develop a new chemical vocabulary that includes such terms as "moles," "molarity," "pH," and "equilibrium." We will describe the mathematical relationships that connect these terms. By learning the names, chemical structures, and properties of acids and bases, we will understand chefs' choice of ingredients to transform their foods, and we will have the tools to understand these processes.

Like temperature, pressure is an external factor that chefs must be aware of and can manipulate when transforming foods. In **Section 2**, we first consider the pressure from the atmosphere and how natural variations in that pressure due to elevation changes change the boiling point of a liquid and necessitate different cooking times and temperatures. What is the science of high-pressure cooking used to cook foods more rapidly and sterilize and preserve foods? What about new techniques of cooking at low pressures, or sous vide cooking? How does this method of cooking preserve flavor and ensure food does not get overcooked?

Finally, in **Section 3**, we will explore how microbes can transform food in an ancient process known as fermentation. By harnessing the power of these microbes, we can change flour into bread, grapes into wine, grains into beer, and milk into cheese. In each case, adding bacteria or yeasts to the raw materials results in more stable, digestible, and delicious products.

Two lab experiments based on these techniques can be found in Part IV—Performing.

SECTION 1: CHANGING PH AS A MEANS OF PREPARING FOOD

What Is the pH Scale?

Understanding how changes in pH can cook food requires developing a chemical vocabulary for acids and bases—most notably, understanding the pH scale and how it measures a solution's acidity or basicity. The first principle to be made explicit is that this entire discussion assumes that the changes will happen in water, or using chemical vocabulary, in an aqueous solution. Since many foods we eat or drink are mixtures of macro- and micronutrients in water, this is not an unreasonable assumption. Chemists call the molecules that dissolve in water the solute, the water itself is called the solvent, and the mixture of the two is called the solution. Knowing the strength of an acid might be necessary in deciding, for example, when a drink such as kombucha is sufficiently fermented that it is safe for consumption. Kombucha is a drink made from a sweetened tea to which particular microbes are added to produce fermentation. As the solution ferments, microbes transform the sugar into an acid that releases hydrogen ions, or H^+, into it. The strength of the acid, or acidity, is directly related to the number of H^+ ions in the kombucha. The more H^+, the higher the acidity. Only when the acidity reaches a certain level can you be sure that any pathogenic bacteria have died and that the kombucha is ready to drink. Determining the amount of H^+ solute per volume of solution, or in chemical vocabulary, the concentration of H^+, allows us to know the strength of an acid.

We measure concentration in many ways: grams per milliliter, ounces per quart, parts per million (ppm), or, for our purposes, moles per liter or molarity. The molarity of a solution is given the symbol M and is indicated by square brackets around the atom, ion, or molecule. In considering acids and bases, our interest is in measuring the concentration of protons, $[H^+]$, and hydroxide ions, $[OH^-]$.

Wait, you ask, what in the world is a mole besides a fuzzy little rodent? Since atoms and molecules are tiny and invisible to the naked eye, it takes many, many of them to be visible. A single grain of sugar contains 1.09954×10^{18}, or about 1.1 quintillion, sugar molecules. Since we need to use such large numbers when counting the number of atoms and molecules around us, and since properties like mass of these tiny things are so very small (1 molecule of sugar has a mass of 5.68×10^{-22} g,) chemists agreed to use a new term, a mole. One mole of atoms or molecules is defined as 6.02×10^{23} particles. One mole of these tiny particles is something that we can see and weigh; one mole of sugar has a mass of 342 g, or 1⅔ cups. Think about how many particles that number represents—almost a trillion trillion atoms or molecules! Defining this unit of measurement allows us to have a convenient unit to measure these tiny things.

The very large number, 6.02×10^{23}, called Avogadro's number, is not arbitrary. It is the number of particles you would find in a certain mass of each element or molecule that is known as the molar mass. For example, the molecular formula for water is H_2O, with two hydrogen atoms and one oxygen atom. Water has a molar mass of 18.0 g/mole, which one can calculate from the periodic table by adding the listed mass of hydrogen (times two) and oxygen. There will be an Avogadro's number, or 1 mole, of molecules in 18.0 g water. If this seems bizarre, remember that around 1990, a new company named Starbucks introduced three new terms for the volumes of their coffee: tall is 12 ounces, grande is 16 ounces, and venti is 24 ounces. Arbitrary? It may be that these were the volumes of the paper cups available to the restaurant—and small, medium, and large were just not trendy terms. Four million customers at Starbucks each day order their coffee using words such as tall, vente, and grande—showing that the terms have stuck and become part of the Starbucks lexicon. Counting using moles rather than molecules or atoms allows us to count something we can touch, see, and weigh.

Now we move on to determining the concentration of an aqueous solution using moles per liter. Suppose we could count the number of solute particles. If we divide that number by Avogadro's number, we have determined the number of moles of solute. We can also measure the volume of the solution (in liters). In that case, dividing the two numbers, moles/liter, gives you the solution's molarity, M.

While we rarely calculate the molarity of solutions in the kitchen, we could. For example, you might like one teaspoon (1 tsp) of sugar in your coffee each morning. One teaspoon of sugar weighs 5 grams. A quick calculation or online consultation gives you the molar mass of sugar ($C_{12}H_{22}O_{11}$), which is 342 g/mol. We can find the number of moles by dividing the number of grams (5.00 g) by the molar mass (342 g): there are 0.0146, or 1.46×10^{-2}, moles of sugar in 1 tsp. Next we need to know the volume of coffee. A Starbucks tall is 12 ounces. By converting ounces to liters, we find that the volume of coffee is 0.355 liters. By dividing the moles (0.0146 moles) by the volume (0.355 liters), we calculate that the molarity of sugar in your sweetened coffee is 0.0411 M. Congratulations, you have unlocked the power of molarity calculations!

Sometimes the molecules, atoms, or ions have properties you can measure once in solution, such as salinity or color. The charge on the H^+ ion affects the passage of electricity, and the conductance through the solution can be measured. A special conductance meter, called a pH meter, does just that: it measures the degree to which the solution conducts electrical charge. The term "pH" is a concentration term specific to H^+. Mathematically, pH is equal to $-\log [H^+]$. Remember that when the ion or molecule is shown in square brackets, $[H^+]$, the term refers to the molarity (M = moles/liter), here the molarity of those H^+ that give rise to acidity. Please note that the negative sign before the log term is essential. It is the reason higher $[H^+]$ results in lower pH. The higher the $[H^+]$, the higher the log $[H^+]$, but with a negative sign in front, that means that the lower the $-\log [H^+]$, the lower the pH.

Acidic foods have high concentrations of hydrogen ions, or $[H^+]$, and the higher the $[H^+]$, the stronger the acid. Basic foods have high concentrations of hydroxide ions, or $[OH^-]$, and the higher the $[OH^-]$, the stronger the base. Figure 4.1 shows a pH scale for common foods, the $[H^+]$ and $[OH^-]$ at each pH, and images of common foods that we consider acidic or basic. Coca-Cola, lemons, and vinegar are all acidic. Lemons taste sour because their juice acts as an acid capable of transforming foods.* Spring water is often close to pH 7 and is considered neutral. The alkaline or alkaline ash diet is one rich in vegetables and grains, as shown in Figure 4.1. It is reported to help with weight loss, prevent cancer, and maintain health. A problematic premise of this diet is that eating alkaline foods will create an alkaline environment in your body—which is false. First, the pH varies widely throughout the body for good and natural reasons: from the acid bath of your stomach (pH < 3.5) to the slightly alkaline but *highly* regulated pH of your blood (pH 7.4) to the alkaline bath of your intestines (pH 8). The lungs and kidneys regulate the pH of blood, and the kidneys regulate the pH of urine. It is *always* best to eat a balanced diet that includes lots of vegetables, fruits, nuts, and legumes, not because they are alkaline, but because they provide minerals, fiber, and other healthy nutrients.

The Power of Hydrogen: A Mathematical Interlude

A good friend of mine refers to "pH" as the "power of hydrogen," which is a great way to remember the mathematical relationship between $[H^+]$ and pH, $[H^+] = 10^{-pH}$, and to remind us of the power of acids, or specifically $[H^+]$, in

* Some tables list lemons and limes as alkaline foods based on their effect on the pH of the urine. The body metabolizes some of the minerals in lemons and limes to produce this result. In food chemistry, we still classify lemons and limes as acidic.

ACIDITY AND BASICITY OF COMMON FOODS

	ACIDIC				NEUTRAL						BASIC			
pH	1	2	3	4	5	6	7	8	9	10	11	12	13	14
MOLARITY (moles/liter)														
$[H^+]$	10^{-1}	10^{-2}	10^{-3}	10^{-4}	10^{-5}	10^{-6}	10^{-7}	10^{-8}	10^{-9}	10^{-10}	10^{-11}	10^{-12}	10^{-13}	10^{-14}
$[OH^-]$	10^{-13}	10^{-12}	10^{-11}	10^{-10}	10^{-9}	10^{-8}	10^{-7}	10^{-6}	10^{-5}	10^{-4}	10^{-3}	10^{-2}	10^{-1}	10^{-0}

Figure 4.1 Acidity and Basicity of Common Foods: pH Scale, $[H^+]$ and $[OH^-]$

transforming foods. If you need a little help with this mathematical rearrangement, remember we defined $pH = -\log [H^+]$, or rearranging, $-pH = \log [H^+]$, then taking the antilog of both sides means that $10^{-pH} = [H^+]$, a concentration term.

Of course, acids have a significant $[H^+]$, but even plain water has a small quantity of $[H^+]$. Where does that $[H^+]$ come from? All by itself, with no heating, no catalysts, and no electrolysis, a tiny fraction of the total number of H_2O molecules spontaneously dissociate to $[H^+]$ and $[OH^-]$ according to the chemical reaction equation

$$H_2O \rightleftarrows H^+ + OH^- \qquad Keq = 1.0 \times 10^{-14}$$

The reactant is H_2O, and the products are H^+ and OH^-. The $\rightleftarrows$ symbol tells us that this reaction goes in both directions; we say it is in equilibrium. Keq is the equilibrium constant whose magnitude indicates the extent of the reaction. For this reaction, the Keq measures the concentration of products divided by the concentration of reactants. Large values for Keq mean the amount of product made is large, and we say that products are favored. Small values for Keq mean reactants are favored.

A familiar example may help illustrate the idea of equilibrium. Imagine a busy marketplace crowded with people. Shoppers are constantly moving in and out of different shops. The forward reaction represents the shoppers entering the shops, and the reverse reaction represents those leaving. The number of people inside the shops remains relatively constant, although individuals still move in and out. This dynamic yet balanced situation mirrors the concept of chemical equilibrium, where the concentrations of reactants and products remain constant and can be represented by the equilibrium constant, Keq. In contrast, individual molecules continue to react and form products while others are converted back into reactants.

Another analogy might help explain the specific water deprotonation reaction above. A party is held, and many guests are invited. It's cold outside, so all the guests (H_2O) arrive with their jackets (H^+) on. Inside, some guests remove their jackets, representing the forward reaction that generates H^+ and OH^-. Here OH^- represents those guests who are now without jackets. After a while, some of these guests get cold and put their jackets back on, the molecular equivalent of the reverse reaction. Meanwhile, others who initially kept their jackets on decide to remove them. There is a constant back and forth of guests taking jackets off and putting them back on, of forward and backward reactions. Once the party gets under way, the overall number of guests with jackets on and those with jackets off remains constant, although the identities do not remain constant. We say the room is at equilibrium, and the equilibrium constant, Keq, represents the ratio of products to reactants or the extent of the reaction.

The Keq for water dissociation is 10^{-14} M. You would agree that 0.00000000000001 is a tiny number; the extent of the reaction is minimal, the amount of product is minuscule, and most of the water molecules do not dissociate. If they were at a party, this would mean that almost all the H_2O molecules kept their jackets on. It is also true that you get one OH^- for every H^+ that comes from water. When a solution has $[H^+] = [OH^-]$, the solution is neutral, and neutral solutions have a pH of 7. At pH = 7, we can calculate that the $[H^+]$ = 0.00000010 M, or 1.0×10^{-7} M. Neutral foods, such as those containing starches and sugars, have a pH of about 7. Remember, at pH 7, the $[H^+] = [OH^-] = 1.0 \times 10^{-7}$ M, a tiny number. Not only are these numbers small in neutral foods, but the acidic H^+ and the basic OH^- are the same: the food is pH balanced.

Back to the Acidity of Foods

Some food molecules release H^+ into solution and are called acids. Foods with an excess of H^+ are acidic and have a pH < 7. Acidic food will taste sour (think of lemon juice or vinegar). The most acidic food regularly consumed is Coca-Cola, with a pH of 2.3. At this pH, $[H^+] = 0.36$ M, and at that high acid concentration, prolonged exposure will take enamel off teeth.

Other food molecules are basic and take H^+ away from water, leaving OH^- behind. Foods with an excess of OH^- have a pH > 7. These are lentils, nuts, legumes, and some vegetables. Another term we use to describe foods with a high pH is "alkaline," which is helpful due to the confusion sometimes created by the ordinary meaning of the word *basic* (fundamental) and the chemical sense of the word *basic* (having a high pH).

Foods treated with the strong base NaOH (lye) can have a pH of up to about 12. At this pH, the $[H^+]$ is tiny, 1.0×10^{-12} M. To determine how much OH^- is present, we define another term, pOH, which is -log $[OH^-]$, and use this helpful relationship: pH + pOH = 14. So if the pH = 12, the pOH = 2. To calculate the $[OH^-]$, we use the relationship $[OH^-] = 10^{-pOH}$ and find the $[OH^-]$ is 0.01 M. Hydroxide ion concentrations higher than this will cause caustic burns in our mouths, so strong bases such as the lye should be used carefully. Other bases used in cooking are weaker than lye and safer. You may be familiar with other strong bases, calcium carbonate ($CaCO_3$) and sodium carbonate (Na_2CO_3), used in various cooking applications. As discussed later, a weak base, baking soda, $NaHCO_3$, is used frequently for leavening foods and browning bread.

In NaOH, the source of the OH^- is right there in the chemical formula, but what about $NaHCO_3$ or baking soda? When $NaHCO_3$ dissolves in water, the ions separate into Na^+ and the bicarbonate ion, HCO_3^{-1}. This ion pulls an H^+ from H_2O, leaving behind an OH^-, which is the source of the alkaline nature of an aqueous solution of baking soda. You might wonder why the other product of that reaction, carbonic acid, H_2CO_3, does not contribute to making the solution acidic. It turns out that H_2CO_3 decomposes rapidly to H_2O and CO_2 gas, the source of the leavening power for nonyeasted cakes and bread.

Sensory Perception of Acids and Bases

Chapter 2 showed that we have unique taste cells for tasting sour, acidic foods with abundant H^+. We learned that the sour taste response results from a channel in the taste buds that allows the passage of H^+. This taste cell is triggered when the concentration of H^+ from your food exceeds the concentration inside the cell. H^+ enters the cell, and the movement of charge across the membrane acts like a wire to send an electrical signal to the nerve cells that connect to our brains. If we taste a lemon with lots of H^+, the brain responds, "WOW! This food is sour."

But what about bases as a culinary category? Do they have a characteristic taste? Do we have a receptor for them as we do for acid? Baking soda is the most common weak base in our kitchens. Try tasting a mixture of baking soda and water by dipping your finger into the mix and putting a bit on your tongue. You will taste a slightly mineral flavor but not much more. You may feel a soapy, slippery texture on your tongue and certainly on your fingers if you rub them together. This sensation results from the base (baking soda) reacting with the natural oils on your fingers, causing the oils to break down to make soap, which is slippery. So while we do not have a specific taste receptor for bases, we know if we taste them through textural properties.

Some foods contain molecules known as alkaloids, which are chemically classified as bases. These naturally occurring compounds exist in foods such as coffee (caffeine) or chocolate (theobromine). Alkaloids are derived from amino acids but are not part of the macronutrient protein group discussed in Chapters 1 and 3. These alkaloids bind to specific bitter receptors in our mouths. At first, most people do not like the taste of bitter compounds. Enjoying them is an acquired taste. Many alkaloids are beneficial for treating human diseases as they are anti-inflammatory, antimicrobial, and antifungal. They are used in drugs that fight cancer and alleviate pain.

pH Indicators

You may notice that the top row of Figure 4.1 is colored red to blue to be consistent with the color changes you might see if you tested the pH with litmus paper. We call such dyes indicators because they "indicate" the pH, though this is not quantitative. For that, you would want to use a pH meter. Litmus paper is simply paper that has been soaked in a pH-sensitive dye initially extracted from a sea lichen that turns red if exposed to acid pH and blue if exposed to basic pH. Today pH paper contains a combination of many different dyes to distinguish slight differences in pH. Many foods will also change color as you change the acidity, which is the source of endless fun in the kitchen. As shown in Figure 4.2, red cabbage is a fantastic pH indicator, turning from red at low pH through the spectrum to green at high pH. Extracts from other foods such as beets, blueberries, cherries, raspberries, grapes, and pomegranates also contain molecules whose color changes when the pH changes. The colors of many spices are also sensitive to pH. Turmeric, for example, changes from yellow in acid solutions to red in basic solutions above pH 7.4.

Introducing the Acids and Bases in the Kitchen

Our pantries and refrigerators are a source of many acidic and basic kitchen chemicals. The extent of their dissociation into ions measures the strength of an acid or base. Strong acids and bases completely dissociate in water into their ions (H^+ and Cl^- for HCl and Na^+ and OH^- for NaOH), and this can cause them to be a bit too reactive for most cooking applications. Weak acids and bases do not completely dissociate in water. Figure 4.3 lists the weak acids and some strong and weak bases found in the kitchen. Vinegar, a common weak acid used in the kitchen, is made up of a solution of acetic acid and water. When acetic acid loses an H^+, the resultant molecule is negatively charged and is called acetate. Generically, weak acids are referred to as HA. When HA loses an H^+ through deprotonation, it leaves behind an A^-. All three species, HA, H^+, and A^-, will coexist at equilibrium. The expression below describes the equilibrium process.

Figure 4.2 pH Indicators from Food

The pH of a solution can be determined by using a pH indicator, such as the juice from red cabbage. A pH indicator is a molecule that changes color at different pHs. Here anthocyanin, the pigment from red cabbage, changes color from red at pH 0 to blue at pH 8 to yellow gold at pH 14. Chemists and chefs can use the color to report the pH of a solution to which the indicator is added.

$$HA \rightleftarrows H^+ + A^-$$

Using the generic terminology, acetic acid is referred to as HA, and the negatively charged acetate ion is referred to as A^-. Weak acids can have many hydrogen atoms. How do you know which ones will dissociate? The hydrogen dissociating from acetic acid is circled in blue in Figure 4.3. This H starts out attached to a carboxylic acid group (–COOH). It is the attachment of the H to an O that is connected to a C with a second O that makes it easy to be removed. The carboxylate group ($-COO^-$) left behind is in a special environment that stabilizes the negative charge because it can be shared between the two oxygens. The rest of the hydrogens do not have the same chemical environment and so stay attached to the parent molecule. We encountered the carboxylic acid chemical group earlier when learning about the structures of fatty acids and amino acids.

Bases enhance many food flavors. The strongest base used in cooking is lye. "Lye" is a chemical term for sodium hydroxide, NaOH, and a generic term for any strong base made by leaching wood ash. The leached wood ash can produce either soda ash (sodium hydroxide, NaOH) or potash (potassium hydroxide,

Name and Chemical Formula	Molecular Structure, Line Drawing (acidic H circled)	Molecular Structure, Ball and Stick
Acids		
Vinegar - acetic acid $C_4O_2H_4$	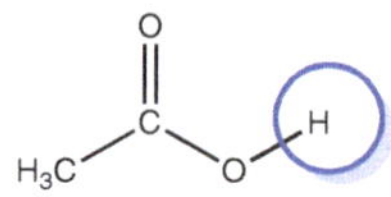	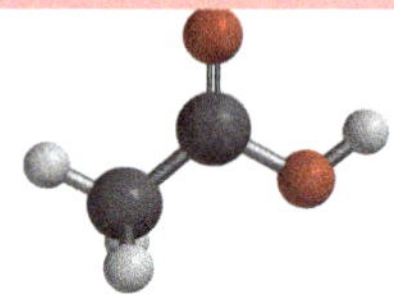
Tartaric Acid (from grapes) $C_4H_6O_6$ Cream of Tartar is the potassium salt $K_2C_4H_4O_6$	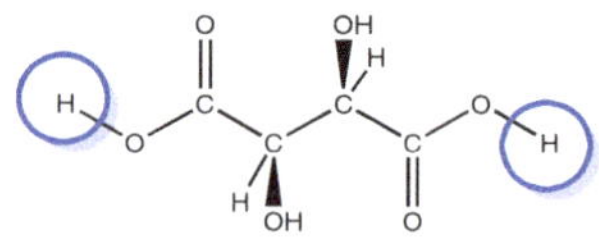	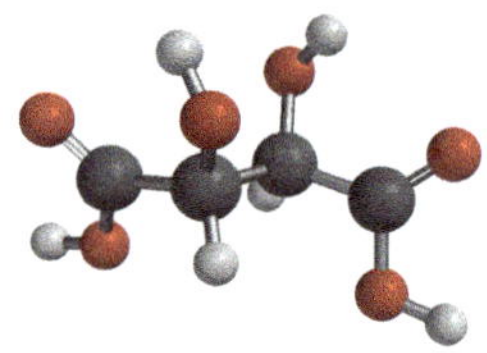
Lemon Juice - citric acid $C_6O_7H_8$	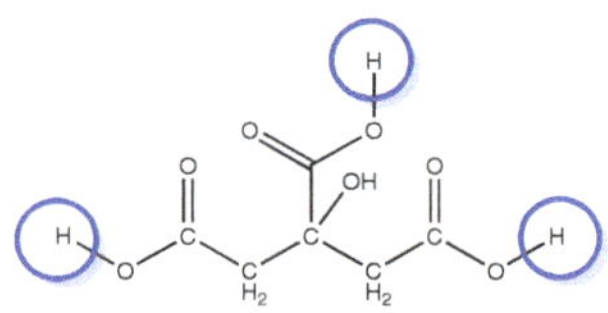	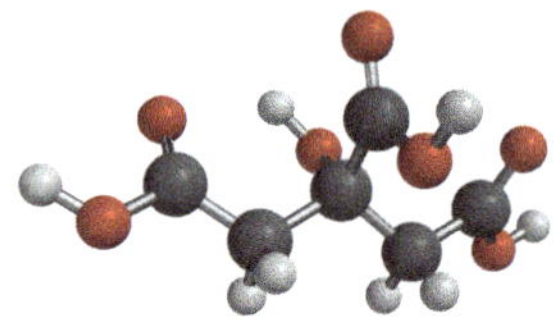
Bases		
Baking Soda sodium bicarbonate $NaHCO_3$	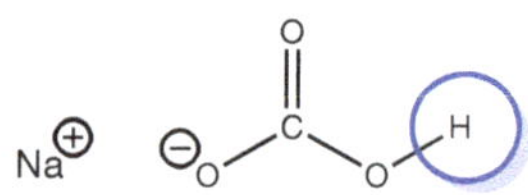	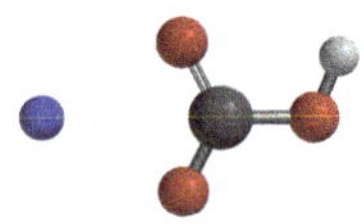
Lime - calcium hydroxide $Ca(OH)_2$	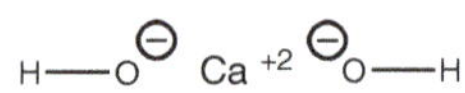	
Lye - sodium hydroxide NaOH	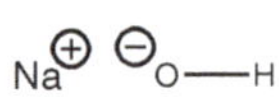	

Figure 4.3 Kitchen Chemicals—Formulas and Molecular Structures of Acids and Bases

KOH). In some recipes, the term "lye" is also used to describe metal carbonates such as calcium carbonate $CaCO_3$ and potassium carbonate K_2CO_3. These metal hydroxides and carbonates are basic and caustic and should be used carefully. Treating cocoa beans with lye produces Dutch chocolate, and treating olives with lye transforms the inedible bitter raw olives into delicious table olives. In both cases, the alkalinity of the base destroys some of the bitter compounds, making the product more palatable.

Baking soda (sodium bicarbonate, $NaHCO_3$) is a weak base used frequently in the kitchen.

Dousing or brushing raw flour dough with an alkaline solution before baking facilitates the browning reactions discussed in Chapter 3 and creates delicious nutty flavors that we love in our pretzels, bagels, and homemade alkaline noodles known as Ramen noodles.

Putting Acids and Bases to Work in the Kitchen

Transforming Seafood with Acid: Ceviché

What is ceviché (pronounced "sah-VEE-chay")? It is any mixture of fish or shellfish "cooked" using an acidic marinade, usually from citrus fruit such as lemon or lime. The presence of chilies, onions, cilantro, and additional ingredients like avocado or coconut can vary depending on the region. Its origin story is murky, with early references to a ceviché-type food traced to its birthplace in Lima, Peru. This country's ceviché is considered the crown jewel in its wonderfully varied cuisine, and July 28 is National Ceviche Day. Today the dish is found throughout Latin America and has become Ecuador's second most popular dish. It is also increasingly popular worldwide. The typical proteins used in Latin America are shrimp, fish, octopus, and black clam. Another root in the ceviché family tree goes back to an ancient Spanish Muslim community, where, some say, it originated before being brought by the Spanish conquistadors to the New World.

When seafood is marinated with vinegar or freshly pressed citrus juice, it undergoes a transformation analogous to cooking meat. Proteins, usually folded at room temperature and neutral pH, unfold when heated or exposed to acid. Salt bridges between positively and negatively charged side groups come apart when the charges change as the pH drops, and the negatively charged side groups get protonated by the acid and become neutral. Hydrogen bonds, which were relatively stable at neutral pH, also become less stable in the presence of additional H^+ in the solution. Even backbone hydrogen bonds that stabilize the helices and sheets become unstable in a protein. Taken together, the proteins in the seafood, especially muscle proteins, unfold, causing the tissue to change from translucent to opaque. This visual change reminds us of how heat will cause the proteins in the white of an egg to unfold, transforming the egg white from a translucent liquid to a white solid. The only factor determining when the seafood is "done" is the time needed for the acid to diffuse into the interior. Is it possible to marinate the seafood too long? Though you might think longer times are safer because the acid will have more time to destroy all the microbes or pathogens in raw seafood, this is not what most chefs recommend. Instead, they say purchasing the best quality seafood (sashimi grade) is better than increasing the marinating time. The acidic liquid should be in contact with the seafood long enough to turn the flesh opaque but not sufficient to turn the fresh seafood rubbery.[1]

Since a key to the success of this dish is using fresh seafood and shellfish and fresh ingredients such as limes, lemons, fruits, and vegetables, ceviché is a dish that varies tremendously depending on whatever is fresh from region to region. In Peru, ceviché dishes feature chunks of cured fish, limes, onions, coriander, and other spices. It is served flanked by wedges of potato or corn. Tiradito is a ceviché from Ecuador that uses fresh fish cut into sashimi-like slices. The fish is cured with sour lemons and plated with local Andean corn or popcorn, a popular side dish.

Chef Douglas Rodrigues, known for popularizing Nuevo Latino cuisine, became a true believer in ceviché during a trip to Ecuador, where he dined at La Lojanita, the mecca.[2] There Chef Rodrigues found that it was not unusual to see fifteen different cevichés, ranging from merluza (hake) with lime and jalapeños to lobster chunks paired with bitter orange and cilantro to buttery tuna with lemon and salty corn nuts. In New York City, Rodrigues created signature ceviché dishes at his two famous restaurants, Patria and Chicama. One is Ecuadorian shrimp ceviché with a sauce of roasted tomatoes and jalapeños and a red bell pepper puree with sweet orange and lime juice. Another dish, Peruvian black ceviche, features mixed seafood flavored with squid ink.

Crossing the globe, in Tahiti, one can find poisson cru. In this local ceviché variant, a sauce of coconut milk, cucumbers, and tomatoes accompanies the lime juice–cured seafood (tuna, prawns, or crab). The unique marinade sets it apart from many of the South American dishes. In Thailand, marinades made with Thai basil and kaffir lime or lychees, soy sauce, and palm sugar accompany the dish. In Fiji, the featured local flavors for ceviché, or Kokoda, are coconut, lime, and chili. Figure 4.4 features a fusion ceviché dish inspired by these flavors.

Transforming Milk with Acid: Ricotta

Making ricotta cheese and fermenting milk to make yogurt both use acids to sour the milk and cause a transformation in the structure of the milk protein casein. "Ri-cotta," or twice-cooked cheese, originates in the kitchens of Italy, where frugal households would have processed milk to extract its curds (milk solids) from the whey not once but twice. Today a favorite recipe for ricotta

Figure 4.4 Foods Can Be Transformed Using Acid or Base

Acids from citrus fruit transform raw seafood into delicious ceviché, as shown in (a). Meanwhile, bases such as sodium hydroxide or lye can transform raw fish into the Scandinavian dish lutefisk, as shown in (b) or raw eggs into century eggs, a Chinese-based culinary delicacy, as shown in (c).

involves heating a milk-cream mixture to just 93.3°C, turning off the heat, and adding lemon juice while stirring until the appearance of a separation of the yellowish watery whey from the fluffy white curd. Adding the lemon juice causes the pH of the milk to drop from about 6.8 to about 5.4. This drop in pH causes the milk's casein proteins to denature and then aggregate to form curds. Lemon juice is added until the separation is complete. At this point, all the casein has unfolded and begun to form curds. Letting the warm milk–lemon juice mixture sit for about 30 minutes allows the curds to grow larger, making them more easily separated. The milky substance left behind, the whey, is full of nutritionally valuable material and can be used instead of water in bread, soups, or smoothies.

Try It Yourself! Instructions for preparing some delicious ricotta can be found in **Experiment 3: Transforming Milk to Cheese and Yogurt** in the culinary explorations in Part IV.

Transforming Foods with Alkaline Solutions

Cooking techniques that use base are much more limited. One method that most home chefs are familiar with uses baking soda, a reliable alkaline substance we know can react with acids to make bubbles that leaven our pancakes or brown our pretzels and bagels. Some cooking methods use alkaline solutions to create unique culinary preparations. One example is lutefisk (literally, "lye fish"). Lutefisk, shown in Figure 4.4, is a Nordic delicacy and is prepared by rehydrating a dried whitefish over days, first in cold water and then in a lye solution to "lute" it—at which point the pH reaches 13 or 14.[3] The original source of the alkaline solution was likely a mixture of lye from wood ash and $Ca(OH)_2$ from lime. As the dried fish rehydrates in the alkaline solution, it swells as the proteins unfold. The alkaline mixture is so strong that it begins to react with protein side groups, causing the fish to become gelatinous. The luted fish is further processed by rinsing and soaking to dilute the lye until the pH reaches about 11. The delicate final product is then cooked gently in water. No traditional Nordic Christmas celebration is complete without lutefisk served with potatoes, mashed peas, and bacon. One colorful origin story for lutefisk suggests that invading Vikings set fire to a coastal village, burning down homes and the village's food stocks. Fish drying on birch bark poles fell into the extremely basic ash created by the burning wood. Later, starving villagers discovered the ash-soaked fish and found that it was pretty tasty when rinsed with enough water.

Bases are also crucial to some browning reactions, especially for bread and pasta. Pretzels require a base wash to get their brown-golden color. If you do not have access to food-grade lye, you can bake baking soda to get a similar result, though your pretzel will not be as crispy and will be lighter and less shiny. Harold McGee recommends spreading a layer of baking soda on an aluminum foil–lined tray and baking it for one hour in an oven at 135°C (275°F).[4] The baked baking soda can be stored in an airtight container. To make pretzels, add 1⅓ cup of baked baking soda to 1 quart of warm water. Immerse the pretzels in the wash for 3 to 4 minutes and then rinse them in a large bowl of plain water before baking. They will come out of the oven toasty brown. Bagels are another bread dough treated with an alkaline wash.

Why is the base wash used in browning bread? You may remember the Maillard reaction described in Chapter 3, where a lysine side group of a protein reacts with sugar and nucleic acid molecules in a complex temperature-dependent way to create delicious roasted, toasted flavors. These reactions require the lysine group to be deprotonated; a basic environment does just that. In flour-based products, the main protein of interest is gluten, the protein most affected by the base. The acidity of some bread, such as sourdough, leaves the lysine groups of gluten protonated, reducing browning. Some bread products, such as pretzels and bagels, take their signature flavor from the brown crust. Bakers can achieve this by dipping the raw dough in a solution of baking soda and water before baking, allowing those lysine residues on the surface of the product to deprotonate and then with the addition of heat, creating those deliciously browned crusts we love so much.

Alkaline pasta is a culinary creation that takes advantage of the reaction between the gluten in flour and a base. Many Sichuan restaurants feature these noodles. The slippery, elastic texture and tawny brown color set them apart from other noodle dishes. To make the noodles, one starts with a noodle dough made from semolina flour, salt, and water with a small amount of added lye (Kansui). The dough is then shaped into noodles. Kansui is a food-grade potassium carbonate solution available in most Asian food stores.

Century eggs are another Asian culinary delicacy in which eggs from ducks, chickens, or quail are covered with clay, wood ash, salt, lye, and rice husks and buried for weeks to months. Over time, the alkaline materials transform the egg proteins to create a delicacy in which the only feature of the egg that remains intact is the shape, as shown in Figure 4.4. The yolk has turned all shades of green, the white has metamorphized into a clear amber gel with small tracks of silver, and the taste is creamy, pungent, and delicious when paired with soy sauce, a drizzle of sesame sauce, some scallions, and firm tofu.

SECTION 2: CHANGING PRESSURE AS A MEANS OF PREPARING FOOD

We exist surrounded by air: a mixture of gases with approximately 70% nitrogen, 20% oxygen, and lesser amounts of water vapor, argon, carbon dioxide, and methane and trace amounts of other gases and pollutants. Those of us at sea level live underneath a column of air that extends upwards to around 10,000 km—and the mass of all this air swirling above and around us creates a compression or pressure of 14.7 lbs/ft^2, which scientists define as 1.0 atmosphere (atm) of pressure. However, since we have always lived with this pressure, our bodies are used to it, and we do not consciously notice it—that is, unless something happens to change it. Ascend to a high mountain pass or go up in an airplane, and your ears will pop as your body struggles to readjust to the adjusted lower pressure.

How do gases create this pressure? Any material in the gas phase contains atoms or molecules that are in constant random motion. Because they are moving, they have kinetic energy. Each time a gas particle strikes the walls of its container, or for our example here, collides with our skin or eardrums, a tiny collision impact is felt. Now imagine the billions of air molecules surrounding us and constantly banging against us from all directions. The pressure we feel will depend on the number of particles in the air around us. The eardrum's sensitivity to the impact of these air particles is how we hear. The thin membrane covering the eardrum's opening acts like the head of a drum and will vibrate as a sound wave hits it. When traveling from sea level, at a lower altitude and higher pressure, up and over a mountain pass, at higher altitude and lower pressure, you feel your ears POP. The physical sensation and sound happen as the higher internal pressure of your eardrum must adjust to the lower external pressure.

Pressure also depends on temperature. As the temperature rises, gas particles will have greater and greater speeds, and their impact on the surrounding walls will become more and more forceful, and the pressure will increase. The same principle causes tire pressure to rise on a warm day or drop on a cold winter morning.

When a gas is in equilibrium with a liquid, such as water vapor and water liquid in a pot or kettle, the gas exerts what we call a vapor pressure on the

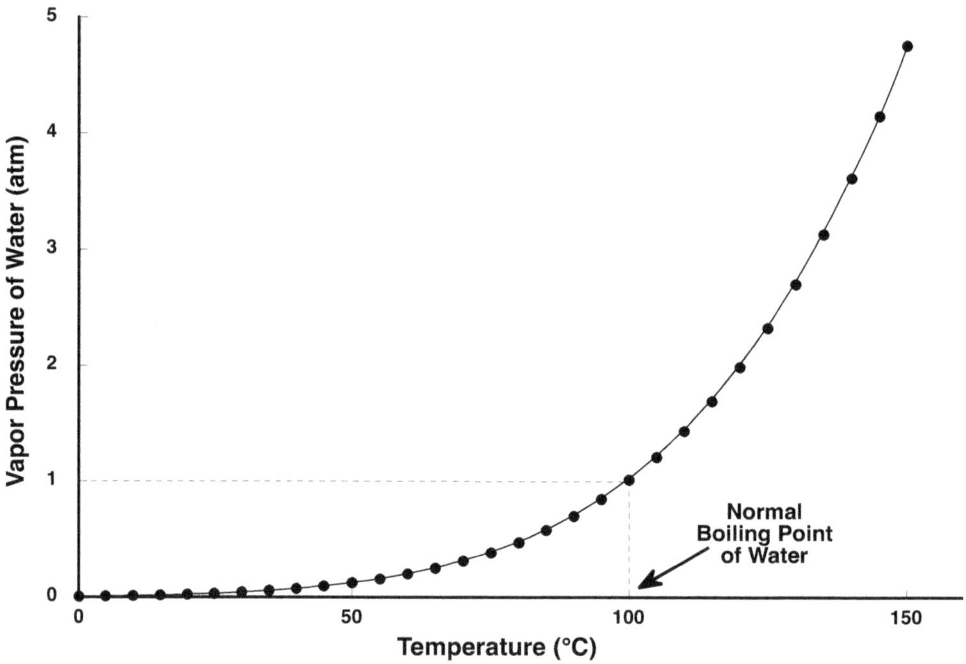

Figure 4.5 Temperature Dependence of the Vapor Pressure of a Gas

As the temperature increases, more molecules of the liquid evaporate into the gas phase, increasing the pressure of the vapor on the surface of the liquid. When the vapor pressure equals the atmospheric pressure, the liquid will boil. Here we see that for water, the temperature at which water vapor pressure equals 1 atm is 100°C, the boiling point of water at sea level.

surface of the liquid. You might remember vapor pressure in the discussion of odors in Chapter 2. At that time, we said molecules called odorants have a high vapor pressure and can leave the liquid or solid phase and enter the gas phase, where they eventually diffuse through the air to your nose to be smelled. Water has no odor, but we know it can also exist as a liquid and a gas (vapor). At room temperature, the vapor pressure of water is about 0.03 atm. As you heat water in a kettle, more and more molecules have enough energy to escape the bonds of the liquid and enter the gas state, so the vapor pressure of the water increases, as shown in the graph in Figure 4.5.

When the temperature reaches 100°C, water vapor pressure equals 1 atm. If you are at sea level, this pressure equals the atmospheric pressure, and the water begins to boil. Those vaporized water molecules in the kettle become steam and exert pressure on the kettle's lid; as the steam escapes, you hear the kettle whistling. Water will boil at a lower temperature if you are at a high altitude with low atmospheric pressure. If you seal water in a closed container, such as a pressure cooker, and add heat, the water vapor is trapped and cannot escape into the surroundings, and the pressure in the pot rises due to the increased number and kinetic energy of the gas particles in the vapor phase. Because of that, the boiling temperature will also rise, and the water and steam will become superheated.

In this section, we describe several ways that pressure is important in cooking. First, we describe two ways pressure can be manipulated to produce desirable culinary results: low-pressure and high-pressure cooking. Both cooking techniques have unique applications, histories, and equipment. Low-pressure, or sous vide, cooking involves removing the air from food in a pouch that is then sealed and submerged in a water bath at an elevated temperature. Cooking without air reduces food degradation due to oxidation and enables more control over the heating process. With high-pressure cooking, heat is added to food in a tightly closed pot containing water. The heat causes a pressure buildup in the air-steam mixture. The boiling point increases at higher pressures, facilitating faster cooking. Since bacteria cannot live at these high temperatures and pressures, pressure cooking also ensures a level of sanitation that helps preserve foods. Then we consider adaptations that need to be made when cooking at high altitudes.

Cooking at Low Pressure: Sous Vide

Sous vide cooking is an increasingly popular method of food preparation that incorporates low-pressure technology. Though many believe this is a relatively new cooking method, it was first developed in 1968 by a retired army colonel, Ambrose McGuckian, who oversaw hospital food service.[5,6] Patients complained that the hospital food was inedible and, therefore, a barrier to their recovery. Funds were limited, so McGuckian explored a method of sealing the food in plastic pouches and immersing it in a heated water bath. He found that this would accentuate the rich flavors of the food. Food cooked by this method could be kept refrigerated for up to sixty days without spoiling and minimized food waste. This relatively simple idea was at the same time being explored by culinary leaders in France, and by the mid-1970s, the technique was being used in fine restaurants, with its new name, "sous vide."

The process involves placing raw food in a food-safe plastic bag with or without a marinade or brine and then removing the air using a vacuum sealer or the water displacement method. In the water displacement method, the air is forced out of the bag with the raw food by gradually and carefully lowering the open bag in water using the outside water to push the air out of the bag. The bag must be sealed once all the air has been pushed out. While rolls of specially formulated plastic that can be heat-sealed are available, simple zip-lock bags work fine. The food is then placed in a constant-temperature circulating water bath at a precise temperature and heated at that constant temperature until the food is done. Sous vide immersion circulators that accurately heat, circulate, and maintain precise temperatures of the water bath are now available to the home chef at prices equivalent to a toaster oven or stand mixer. This gentle method of heating prevents overheating or overcooking and helps the food maintain the moisture necessary for the best texture. It is not a fast-cooking method. The heating process can take an hour or more to ensure thorough heating and the destruction of any food pathogens. At the same time, food can be held entirely cooked for several hours at elevated temperatures without ill effects—a significant advantage in catering, restaurants, and other food service establishments. Just before plating and serving, a quick sear over a hot flame produces the Maillard browning that adds flavor.

The air removed from the bag is a mixture of gases, and while it is predominantly the inert gas nitrogen, it also contains about 20% oxygen. By removing the air from the food bag, you have removed the most crucial molecule responsible for oxidation, molecular oxygen or O_2. To understand oxidation, you must remember that all atoms are composed of three subatomic particles: protons, neutrons, and electrons. Atoms have no charge when they are in their elemental states. We can ignore the uncharged neutrons for this conversation but must consider the positively charged protons and negatively charged electrons. As we remember from the previous chapter, oxidation happens when the atoms or molecules in food lose electrons, to another atom, usually oxygen, an element hungry for electrons. Food degradation is the result. When we slice fruits or vegetables, the browning results from reactions of the exposed fruit with molecular oxygen.

How safe is it to use plastic bags in sous vide cooking? Since the food is prepared in plastic bags, sometimes with acidic marinades, and then heated for extended periods, some consumers worry about the health risks of eating foods prepared by this method. To be as safe as possible, most plastic products available for sous vide cooking use food-safe plastics such as those made from polypropylene and avoid unsafe plastics such as polyvinylchloride that can leach toxic monomers or polycarbonate that can slowly release bisphenol A (BPA), a structural component in the manufacture of polycarbonates. Since BPA has been implicated in inducing estrogenic-like stimulation in human cells, in 2012, the FDA recommended limiting its use in food production, and BPA-free plastics that avoid this potential toxin are now available.[7,8] BPA-free polypropylene is one of the safest food-grade plastics currently in use, but even this plastic cannot claim to be 100% safe. Recent studies have shown it has problems similar to those of the other plastics but at a much lower level—and we know that it provides the safest sous vide plastic currently available.[9]

Try It Yourself! Explore how cooking foods in a vacuum can improve flavor and texture by following the instructions in **Experiment 4: Sous Vide—Cooking at Low Pressure without Oxygen**, in culinary explorations in Part IV.

Cooking at High Pressure / Pressure Cooking

Water and steam are heated in a pressure cooker above their normal boiling point. The increased pressure and higher temperatures mean the food is cooked much faster. Transforming food with this pressure-cooking method is certainly not a recent invention, as the first pressure cooker traces its origin to the seventeenth century when the mathematician and researcher Denis Papin invented a cooking device called a steam digester.[10] The sturdy device could be filled with food, such as roast mutton in broth, and had a lid that could be screwed down. When the digester was heated over a fire, steam got trapped inside, which led to an increase in temperature and pressure, thus resulting in faster cooking. A release valve allowed the device to be opened safely after cooking. In the mid-twentieth century, the first-generation pressure cookers became common kitchen appliances; they were used for canning foods, sterilizing baby bottles containing formula, and cooking certain foods more quickly.[11]

Today, modern multi-cookers such as the Instant Pot take these innovations one step further, incorporating an electric heater, timer, preprogrammed settings, and, with most recent models, multiple cooking options such as sautéing, steaming, sterilizing, rice cooking, and slow cooking. Traditional methods of cooking dishes such as risotto are labor-intensive and prolonged. Multi-cookers can transform cooking risotto into a simple two-step process in minutes.

Adjustments for Cooking at High Altitudes

Those who live at high altitudes must adjust recipes and boxed directions when cooking and baking. For example, in the mile-high city of Denver, Colorado, the air pressure is typically around 0.83 atm, compared to 1.0 atm at sea level.[12]

It might be helpful to think of atmospheric pressure as a blanket of gaseous particles pressing down on the water's surface. At water's boiling point, the vapor pressure equals the atmospheric pressure. As the atmospheric pressure drops from 1.0 atm at sea level to 0.83 atm in high-altitude cities like Denver, the boiling temperature will drop from 100°C (212°F) to 94°C (198°F). What effect does this have on cooking? The lower temperature means boiling pasta, cooking an egg, or making a béchamel sauce will take longer. Attempts to raise the temperature will not make it cook faster because getting water hotter than 94°C is impossible without a pressure cooker at these altitudes. It is also true that roasting time for meat and poultry will need to be increased, as they are typically about 75% water. It is not unusual to see a high-altitude cooking time increase by 25%.

For this reason, modern pressure cookers such as the Instant Pot are a significant advantage and time saver in high-altitude environments. Air at high altitudes is also much drier, making moisture a big concern in high-altitude kitchens. Frequent basting, sauces, and secure wrapping of foods help preserve a dish's natural moisture and juiciness. Finally, gases like CO_2 needed to leaven cake or bread expand more than at sea level. A recipe might need to be adjusted for the amount of yeast or baking soda added.

SECTION 3: TRANSFORMING FOOD BY FERMENTATION

Fermentation is an ancient process by which larger biological molecules can be broken down into smaller molecules by the action of yeasts or bacteria.[13] Typically, the microorganisms will digest starch or sugar, breaking them down to ethanol, carbon dioxide, and energy—a process called alcoholic fermentation. The microbe uses the energy to live and grow. The chemical reaction catalyzed by yeast enzymes is

$$\text{Glucose } (C_6H_{12}O_6) \rightleftarrows 2 \text{ ethanol } (C_2H_6O) + 2\, CO_2 + \text{energy}$$

While this process has happened in the natural world for billions of years, humans have used fermentation to transform food for at least 10,000 years. Recently, archaeologists recovered evidence from two sites in Georgia in the South Caucasus of early humans' cultivation of grapes and the transformation of the grapes into wine from vessels containing the remains of wine that date to the early Neolithic Period, 6000 BCE.[14]

Fermentation not only breaks down food into more digestible molecules, but it also preserves the food from further spoilage. In the days before refrigeration, this was a most welcome feature. Early human settlements and the cultivation of plants meant crops could be harvested and preserved for times when food was scarce. Humans learned how to ferment grain, fruit, milk, and vegetables to make bread and beer, wine and vinegar, yogurt and cheese, and pickles. The yeast used to make wine (or beer) is a common strain known as Brewer's yeast, or *Saccharomyces cerevisiae*. Whereas the grape juice would spoil in a few days, the fermented wine is stable for long periods and requires no refrigeration. The Speyer Wine Bottle, recovered from a Roman tomb in southwestern Germany in 1867, is estimated to be 1,700 years old and is believed by experts to be still drinkable.[15] Almost one-third of the foods we eat today are produced by fermentation.

When we bake bread, we take advantage of the same alcoholic fermentation processes that make wine: in the presence of yeast, sugar starches break down to alcohol, CO_2, and energy. However, in this case, instead of being released to the surroundings, the CO_2 is trapped in the dough by the gluten proteins. This gas expands as it is heated, creating bubbles that leaven or lift the bread, making it light and delicious. The second fermentation product is alcohol; you may have smelled this in the raw dough before baking. Once in the oven, other natural aromas associated with baking mask the odor of any residual alcohol. One can make bread, such as sourdough bread, by mixing flour and water and taking advantage of our environment's wild yeasts. More often today, bakers use Baker's yeast, a strain of *S. cerevisiae* that is a reliable and quick-acting alternative to wild yeasts.

The yeast responsible for making wine, beer, and bread, *S. cerevisiae*, is a facultative anaerobe. That means that while it prefers to live without molecular oxygen (anaerobically), it can survive in oxygen when it is also in the presence of sugar. Given that our planet, which is 4.5 billion years old, spent its first billion years without any life forms and another billion years with only trace amounts of oxygen in its atmosphere, it is likely that the ancestors of modern-day *S. cerevisiae* evolved to be anaerobic. It is not surprising to learn that ancient Egyptians believed the actions of yeast were a miraculous gift of the gods. It was not until 1860 that Louis Pasteur used scientific methods to study fermentation. He suspected that yeasts were the agents responsible for fermentation and established their central role in alcoholic fermentation. He also demonstrated that yeast cells can live with or without oxygen. Pasteur showed that yeast was indispensable for forming bread's aromas and flavors. Today, *S. cerevisiae* is a model organism used by microbiologists. It is the first eukaryotic organism (organism with a nucleus) to have its complete genome sequenced.[16] Figure 4.6 highlights the many foods and food-related applications that rely on this model organism.

Transforming Milk with Microbes: Fermentation of Yogurt

Yogurt is a fermented food made from milk that traces its history to about 5000 BCE. Regional fermented dairy products, in addition to yogurt, include sour cream, buttermilk, cheese, kefir, lassi, labneh, and ayran—a small sampling of the vibrant array of liquid and solid products, some sweet, some savory, some plain, some adorned. Figure 4.7 shows the structure of the complex milk sugar lactose and its breakdown product, lactic acid, that produces the characteristic "tang" you taste in plain yogurt. The microbial fermentation process that creates yogurt relies on the slow digestion of lactose by the bacteria in the live starter culture (typically, *Lactobacillus bulgaricus* and *Streptococcus thermophilus*). These bacteria work together to continue the fermentation process for four to twelve hours. Given its preference for growing in media with neutral pH and high oxygen content, the *S. thermophilus* kicks off the show. As the sugars and proteins break down, the pH and the oxygen content drop, which are perfect conditions for the growth of *L. bulgaricus*. Once the lactose is gone, fermentation stops. The lactic acid product is a weak acid that causes the pH to drop and produces physical changes in the milk. The milk proteins gradually unfold and then aggregate, forming a material with a soft gel-like matrix. Flavor compounds characteristic of yogurt are also produced during fermentation.

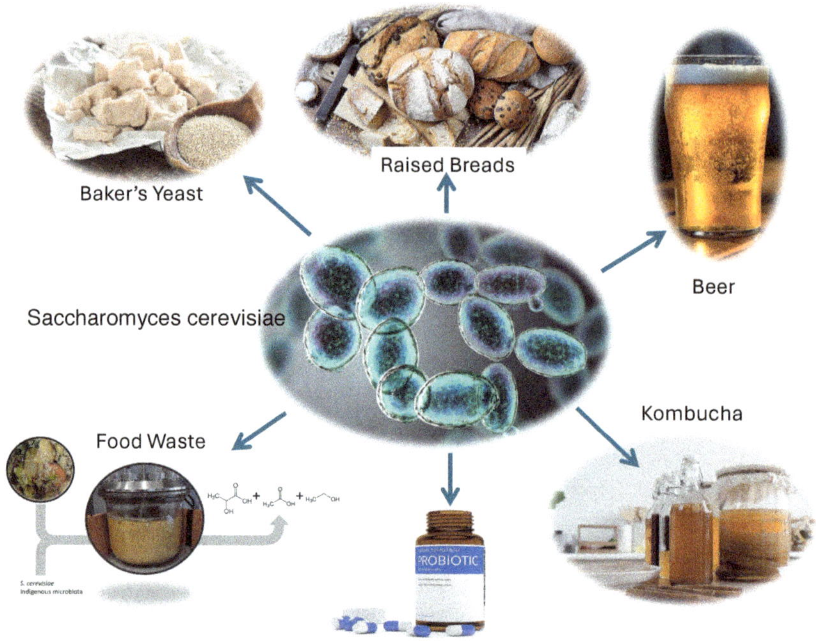

Figure 4.6 The Most Popular Microbe Used in Fermentation Is *Saccharomyces cerevisiae*

This model organism is commonly known as Baker's yeast or Brewer's yeast and is used in baking bread, brewing beers, and other fermented beverages. It also provides additional supplementary nutrition through probiotics and fortified foods and can be used to treat food waste.

Making yogurt starts by heating milk to about 85°C, a temperature sufficient to kill any competing bacteria in the milk but not high enough to denature the protein and form curds. Next, when the milk cools to about 45°C, it is "inoculated" with active cultures from a yogurt starter. The two bacteria in the culture consume the sugar and proteins of the milk, producing the breakdown product lactic acid, and after four to twelve hours, at around 30°–40°C, the yogurt is done.

Yogurt typically has a much lower pH (about 4.6) than ricotta (about 5.4). Remember, if the pH differs by 1, the $[H^+]$ differs by a factor of 10. While each product has been created by the action of a weak acid, in one, the acid was added directly, and in the other, the acid was generated by microbial activity from added cultures. Both are good for you; both are delicious.

Try it yourself! If transforming foods using microbes sounds like fun, try **Experiment 3: Transforming Milk to Cheese and Yogurt**, in culinary explorations in Part IV.

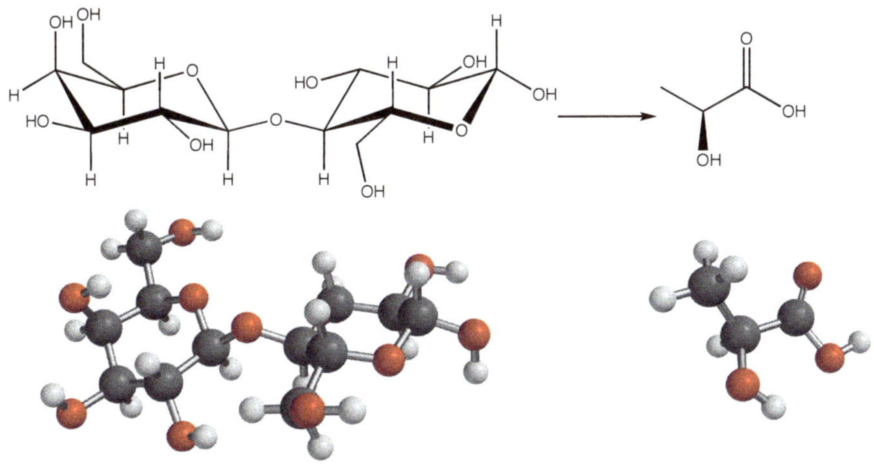

Figure 4.7 Bacteria in Yogurt Converts the Disaccharide Lactose into Lactic Acid

The molecular structures of lactose and the digestion product lactic acid are shown as line drawings and ball-and-stick images. Bacteria added to the warm milk breaks down the disaccharide lactose into simple monosaccharide sugars, which are further broken down and oxidized to lactic acid. The drop in oxygen concentration and pH creates an optimal environment for a more complete bacterial transformation.

Transforming Foods by Fermentation Preserves Them

Foods made by fermentation will often be further processed to ensure the destruction of food pathogens. Various preservation methods have been employed, such as pickling, brining, curing, corning, drying, and smoking. We are most familiar with adding salt, which removes liquid from the food—most commonly, a fruit or vegetable—to make a brine that surrounds the food and keeps it uniformly bathed in a salty solution. This treatment contributes to preserving the food since many pathogens cannot grow in high salt. Pickles, preserved lemons, and olives are foods treated with salt during fermentation. Corned beef is a brisket of beef preserved with salt (the Old English word *corn* was used to describe any material that came in small hard particles like coarse salt) and fermented for several days before cooking. Other foods, such as cacao, coffee, and vanilla, combine fermenting and drying steps to transform bitter compounds into more edible mixtures and then preserve them. Smoking, typically applied to meat or fish, exposes the flesh of these animals to smoke for prolonged periods. This cooking method allows the smoke to flavor the food and kill pathogens that can spoil the meat.

SUMMARY OF CHAPTER 4

In this chapter, we saw how the chemical concepts of pH, acidity, and basicity or alkalinity were useful in understanding how acids and bases can transform foods directly by denaturing proteins. Neutral solutions at pH 7 have $[H^+] = [OH^-]$. Acids produce solutions that are acidic with pH < 7 and the $[H^+] > [OH^-]$. Bases produce solutions that are basic, or alkaline, with pH > 7, and the $[H^+] < [OH^-]$. We saw examples of foods cooked with both acids and bases. When talking about pressure cooking, we learned how our atmosphere exerts pressure on us due to all the gas particles that are constantly moving. We saw that when a liquid and vapor are together, the gas vapor exerts pressure on the surface of the liquid, and when that pressure equals the atmospheric pressure, the liquid will boil. We saw how trapped water vapor in a pot with a sealed lid produces superheated steam that rapidly cooks and purifies food. We also saw that removing the air from food by putting it in a pouch and evacuating it allows foods to be cooked without losing flavor due to oxidations. It also allows food to be prepared ahead of time and held safely at a particular temperature without getting overdone, a real time saver for restaurants. Since pressures vary regionally, these variations will affect cooking. At lower pressures at high altitudes, cooking methods need to be adjusted for the lower boiling point of water and the greater volume of leavening gases in baking. Finally, we learned how fermentation can transform complex carbohydrates and sugars into smaller molecules of ethanol and CO_2. Fermentation has been used for thousands of years to preserve food and provide unique flavors.

Chapter 5 will take us into new territory as we explore the importance of texture in our foods and how to achieve delicious new textures by adding mechanical energy to create stable colloidal mixtures.

Are you interested in trying some of these techniques for yourself? Check out the two lab experiments, **Experiment 3: Transforming Milk into Cheese and Yogurt** and **Experiment 4: Sous Vide**, in culinary explorations in Part IV.

CHAPTER

5 Manipulating Texture

In contrast to the previous two chapters, in which we discussed food transformations resulting from chemical changes, this chapter introduces how we transform foods by changing their phase or texture physically. Before explaining physical transformations, we will review in **Section 1: Perception of Texture** how sound and touch, especially mouthfeel, create the sensory signals by which we perceive texture. Individual preferences, guided by experience and culture, greatly influence our affinity for various food textures. Because mouthfeel is so closely associated with flavor, we must understand the importance of the physical changes necessary to create these transformations. In **Section 2: Mixing the Unmixable**, we will define and understand the science of mixing materials that do not normally mix to create new foods with new textures and mouthfeels. The two mixtures most relevant to textural transformations in food are colloids and suspensions. A suspension is an unstable mixture of particles of one substance dispersed throughout particles of a second substance that separate into separate phases upon standing.

Colloids, on the other hand, do not separate upon standing. Colloidal foods can have physical properties very different from their ingredients. By learning the vocabulary to describe colloidal properties, we will understand how rearranging the molecules and particles can produce these effects. **Section 3: Kitchen Colloids** explores our favorite culinary colloids and suspensions, such as mayonnaise, whipped cream, meringues, gravies, and sauces. **Section 4: Novel Emulsions with Molecular Gastronomy** explores the novel colloids such as foams, éspumas, and spheres introduced by molecular gastronomy.

SECTION 1: PERCEPTION OF TEXTURE

Creaminess, roughness, crunchiness, crispness, stiffness, elasticity, and moisture are just some ways we might describe the "mouthfeel" of a dish. All of these can profoundly affect whether we like it or not. Some people find oatmeal, oysters, okra, or mushrooms repulsive. However, it is not the flavor but the mouthfeel that gives rise to these strong opinions. Similarly, we often describe foods we love using a textural reference—a crisp apple or a creamy dessert. These properties of food are not related to chemical interactions between the food and the taste buds but instead arise from the physical architecture of the food itself.

We all began our culinary journey by using our hands to feed ourselves. Before the invention of utensils, all humans ate with their hands. Today, eating with your hands is an integral part of the culinary culture in certain parts of the world. Some research studies have shown that tactile food exploration with the hands enhances one's appreciation of food.[1] In these cases, the mechanical sensory perception starts with touch. The food might feel sticky, creamy, tough, gritty, or slimy. Bringing the food to the mouth, we take our first bite. Is that first bite of apple crisp and juicy, or is it soft and mushy? We hear the sounds created as our incisors rip apart the individual plant cells and feel the mechanical pressure and resistance our teeth experience as we bite. After that first bite, texture begins to evolve as the food breaks down through the continued chewing action of our molars (mastication). Saliva bathes the tiny particles created through chewing. With its mixture of digestive enzymes, saliva begins to break down the complex food molecules. Starch is broken down into sugar by an important salivary enzyme, amylose. Digestion has begun. Continued chewing creates a wad or bolus of food. Finally, we swallow the food through the muscular action of the tongue. Each step of this process can produce a different texture that we find pleasurable or repulsive. When communicated to the brain and integrated, these signals are combined and interpreted as a texture or mouthfeel.

Think about your favorite food. Now think about your least favorite food—even the thought of tasting it makes you gag. The reasons some foods signal hedonic pleasure and others primal disgust are incredibly complex and culturally and experientially embedded. Charles Darwin believed food aversion was related to evolutionary fitness in that sensitive disgust responses protect humans from eating diseased or rotten foods and thus provide an evolutionary advantage.

SECTION 2: MIXING THE UNMIXABLE

We learned in elementary school that pure substances can be solids that hold their shape, liquids that flow, or gases that fill their container. We also learned that a mixture combines two or more pure substances. But scientists (and chefs) like to be more imaginative than that, and fortunately for us, our culinary world is full of substances that defy the classic definitions. Not only are some solids deformable, such as jello, but some liquids, such as ketchup, are resistant to flow. The kitchen becomes the laboratory where combinations of pure substances create complex mixtures such as dressings, sauces, gravies, and desserts.

Three large categories of mixtures are particularly important in the kitchen: solutions, suspensions, and colloids. Solutions such as salt water are stable combinations of pure substances. The solute, salt, completely dissolves in the solvent, water. Suspensions and colloids behave differently from solutions, sometimes to great advantage when considering texture.

What Is a Suspension?

A suspension is an unstable mixture of two immiscible substances, meaning they do not dissolve. One of the most common suspensions in the kitchen is a simple vinaigrette. To make a vinaigrette, we start by combining oil and vinegar. Since vinegar is about 95% water and 5% acetic acid, it is considered the aqueous layer. The image on the left in Figure 5.1 clearly shows that the oil and water do not mix. The oil layer on the top contains nonpolar triacylglycerides. The water layer on the bottom is primarily polar water molecules. The polar water cannot stably dissolve the nonpolar triacylglycerides. The next step in making a vinaigrette is to vigorously shake the bottle, which adds mechanical energy to shear the oil droplets. The result is an unstable mixture—a suspension of oil droplets in water that persists briefly before spontaneously returning to the phase-separated mixture. Later we will show how adding an emulsifier, like mustard, with vigorous shaking will stabilize the vinaigrette and help keep the two phases mixed, as seen in the second image in Figure 5.1.

What Is a Colloid?

All colloids are stable mixtures and contain a dispersed phase and a continuous phase. To be classified as a colloid, the size of the collection of particles (or drops) in the dispersed phase must be small, that is, between 0.001 micron (about the size of a dust particle) and 1 micron (the diameter of a human hair). We can see drops larger than 1 micron with our naked eyes; a mixture with drops this large is usually unstable and is called a suspension. If the size is smaller than 0.001 micron, the mixture of the two substances appears to be

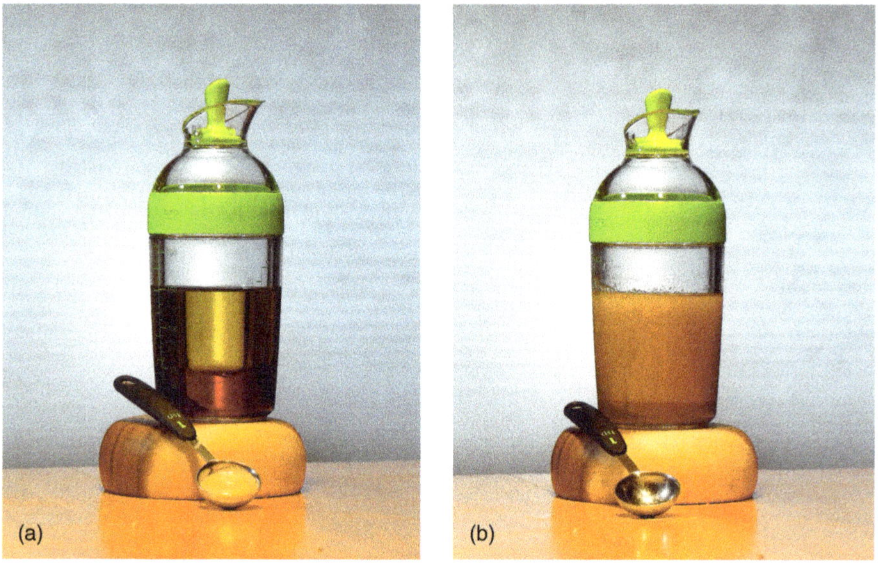

Figure 5.1 Suspension of Oil and Water Can Be Stabilized by Adding Mustard

As a result of their different polarities, oil and water don't mix, as shown in the first image by the separation of red wine vinegar into the bottom layer and olive oil at the top of the cruet. A teaspoon of mustard stands ready. In the second image, the mustard has been added, and the suspension of vinegar and oil has been vigorously shaken, creating an emulsion in which molecules in the mustard surround the oil droplets and hold them in solution.

Table 5–1 Culinary Examples of Colloids

Dispersed Phase	Continuous Phase	Name	Example
Solid	Solid	Sol	Peanut brittle
Liquid	Solid	Emulsion/gel	Cheese, butter
Gas	Solid	Foam	Marshmallow
Solid	Liquid	Sol	Gravy
Liquid	Liquid	Emulsion	Milk, vinaigrette
Gas	Liquid	Foam	Whipped cream
Solid	Gas	Aerosol	Smoke
Liquid	Gas	Aerosol	Fog, mist
Gas	Gas	NA	NA

SOURCE: Modified from Abheetinder Brar, "Colloids," Chemistry LibreTexts, October 2, 2013, https://chem.libretexts.org/Bookshelves/Physical_and_Theoretical_Chemistry_Textbook_Maps/Supplemental_Modules_(Physical_and_Theoretical_Chemistry)/Physical_Properties_of_Matter/Solutions_and_Mixtures/Colloid.

dissolved and is called a solution. Table 5–1 lists several colloidal mixtures and identifies the continuous and dispersed phases.

The vinaigrette shown on the right in Figure 5.1 is an emulsion, a subcategory of colloids in which a liquid is dispersed in another liquid. In this case, oil is the dispersed phase, and water is the continuous phase. The mixture has been stabilized by adding mustard, which acts as an emulsifier.

Other colloids, like jello or jam, are soft solids known as hydrogels. To make jello, you dissolve a powder of gelatin (the dispersed phase) in boiling water (the continuous phase) that is flavored and sweetened. The gelatin, which is animal protein, dissolves at high temperatures but slowly aggregates as it cools to form the colloid with the water. Jam (or, when filtered, jelly) is made by boiling fresh fruit to release the fruit juice that is mostly water, which is the continuous phase. Sugar is added as a sweetener. Pectin, a polysaccharide naturally occurring in fruit, dissolves in the hot mixture and, upon cooling, cross-links to set up a colloid that is easily spreadable on your peanut butter sandwich or breakfast scone. Pectin levels vary in different fruits: quince, apples, and red currants are rich in pectin. Other fruits may need additional pectin to gel correctly.

Properties of Emulsions

Developing a verbal and mathematical vocabulary for describing colloids and emulsions will help us understand them. Emulsions do not form spontaneously but must be shaken, whisked, stirred, or blended. By doing so, we add mechanical energy that shears the drops of the dispersed phase, making them smaller and better able to create the emulsion. One can describe an emulsion by a property known as the volume fraction and given the Greek symbol psi, ϕ.

$$\phi = \text{Volume}_{\text{Dispersed Phase}} / \text{Volume}_{\text{Total}} \qquad \textbf{Equation 5–1}$$

In a stabilized salad dressing, one might use ¾ cup oil to ¼ cup vinegar for a volume ratio of 3:1 and add a small amount of mustard. For the oil and water emulsion described above, where oil is the dispersed phase and water/vinegar/mustard the continuous phase, the volume fraction of the oil, ϕ_{oil}, can be calculated using Equation 5–1:

$$\phi_{\text{oil}} = V_{\text{oil}} / (V_{\text{Total}}) = V_{\text{oil}} / (V_{\text{oil}} + V_{\text{water}})$$
$$\phi_{\text{oil}} = \tfrac{3}{4}\text{ cup} / (\tfrac{3}{4}\text{ cup} + \tfrac{1}{4}\text{ cup}) = \tfrac{3}{4} / 1$$
$$\phi_{\text{oil}} = 0.75$$

Emulsion properties depend on ϕ. As ϕ increases, the emulsion might become thicker. Some colloids undergo a phase change from a fluid to a soft solid once you reach a particular ϕ called the critical volume fraction, ϕ_c. Figure 5.2 shows how the properties of a familiar oil and egg emulsion, mayonnaise, change as the ϕ of oil increases. Slowly add oil to the egg yolk with continual

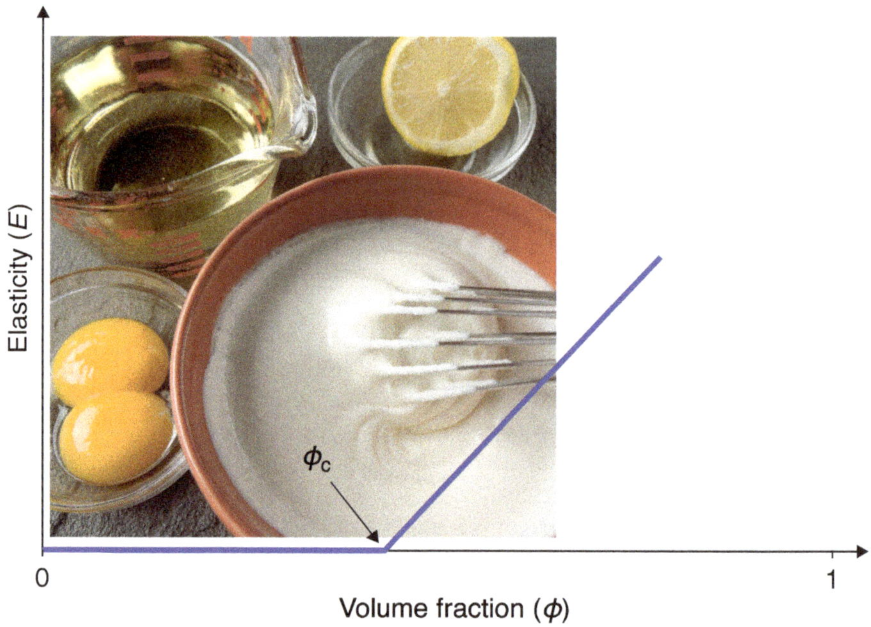

Figure 5.2 Elasticity vs. Volume Fraction for Mayonnaise

Making mayonnaise involves slowly adding oil to egg yolks while vigorously mixing the two liquids. The y-axis measures E, the resistance of the mixture to being deformed elastically when a stress is applied. At the start, the mixture behaves as a liquid, and the E is zero. As the volume fraction of oil increases and the oil is sheared by the action of the whisk into smaller and smaller droplets, the mixture thickens but is still liquid. At a certain point, when ϕ exceeds the critical volume fraction ϕ_c, the mixture transforms into a soft, solid-like substance. The elasticity rises as there is more resistance to deformation.

whisking to make mayonnaise. As with salad dressing, the oil is the dispersed phase, and water is the continuous phase. The water comes from the egg yolk, which has proteins, fats, and, most importantly, lecithin. Once the mixture reaches the critical volume, it transitions to a soft solid. Figure 5.2 shows that this point is reached about halfway along the x-axis when $\phi = \phi_c = 0.5$. Above ϕ_c, you will have mayonnaise with its characteristic soft, solid appearance.

When whipping cream or beating egg whites, we see the phase transition we are looking for with our eyes, but we can also measure it using a particular property of these soft solids known as elasticity. Elasticity, E, measures the mixture's resistance to being deformed when stress is applied. If deformed, say, by poking it with your finger, an elastic solid will spring back to its original shape. In the colloids we see in the kitchen, like jello, whipped cream, egg whites, and mayonnaise, the dispersed phases are all above their ϕ_c. The colloid is elastic; when deformed, it will spring back. The y-axis of Figure 5.2 represents the elasticity of the mayonnaise. We see that below the ϕ_c, the elasticity is zero because there is no resistance to deformation in the fluid. At some point, as we whip more and more oil into the eggs, the ϕ increases above ϕ_c, and the mixture transitions from a liquid to an elastic solid, as evidenced by the sudden onset of elasticity, which then increases as the ϕ increases.

Elasticity, E, is related to other properties of the emulsion, such as the radius of the dispersed drop, R, and the surface tension (represented by the Greek symbol rho, σ), as shown in **Equation 5–2**.

$$E = [(\sigma)(\phi - \phi_c)] / R \qquad \textbf{Equation 5–2}$$

ϕ and ϕ_c are the now familiar volume fractions and the critical volume fraction. If ϕ is above ϕ_c, increasing the surface tension, σ, or decreasing particle size, R, will increase the elasticity. Lowering the temperature or adding a thickening agent such as glycerol, gelatin, or flour increases elasticity by causing an increase in the surface tension of the droplet. The particle size can be decreased by more vigorous whipping, beating, or stirring as mechanical energy shears the drops of the dispersed phase. Vigorously whisking one tablespoon of fat breaks it into more than one billion particles.

Occasionally, the emulsion fails to form or forms but then separates again into two unmixed phases. When this happens, we say the emulsion "breaks." You may have added too much oil or whisked too vigorously, or the temperature is too high or too low. You are left with an unappealing oily-eggy mixture. Some home chefs claim that one magic ingredient for making foolproof mayonnaise is a teaspoon of water or lemon juice added to the egg yolks, ensuring the continuous phase has enough water.[2]

How Do Emulsifiers Keep Colloids from Breaking?

But what about the sauces, dressings, and gravies we hope to use as flowing liquids? First, it is critical to keep ϕ below ϕ_c, something we can do, for example, by adding water to a gravy that gets too thick. Since the surface tension decreases as we increase temperature, we can also warm the mixture to keep it flowing if that is consistent with how we hope to use it. We must also be concerned with keeping the emulsion stable. What do chefs do to keep their emulsions from breaking? Fortunately, nature has provided us with several different types of molecules called emulsifiers.

Emulsifiers are molecules that have a mixed character.[3,4] One part of the molecule is polar, and the other is nonpolar, giving the whole molecule a property called amphiphilicity (liking both). Figure 5.3 illustrates how an emulsifier interacts with both the continuous and dispersed phases, surrounding the dispersed droplets and preventing them from coming together (coalescing), which leads the emulsion to break or separate. The emulsifiers will enable us to produce silky smooth sauces, gravies, and dressings by preventing phase separation. Molecules that act as emulsifiers are derived from all the macronutrient

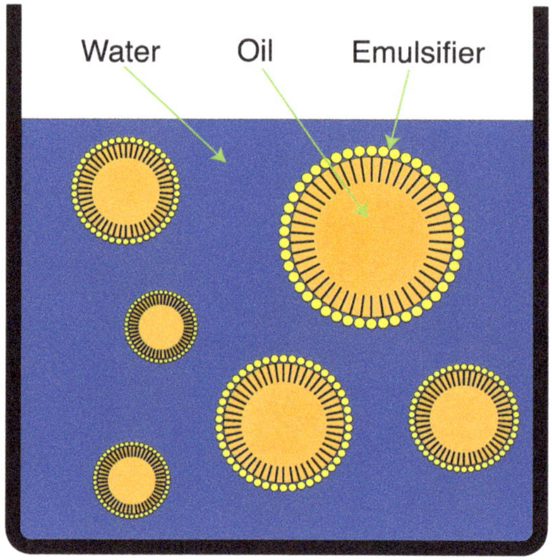

Figure 5.3 How Emulsifiers Mix the Unmixable

Emulsifiers are molecules in which one part of the molecule (round yellow circle) is soluble in the continuous phase (blue background) and the other part of the molecule (black tail attached to circle) is soluble in the dispersed phase (orange-gold drops.) Emulsifiers stabilize emulsions by surrounding droplets of the dispersed phase and preventing them from joining together to coalesce and separate.

groups: proteins, fats, and carbohydrates. Figure 5.4 lists emulsifiers derived from fatty acid derivatives such as lecithin from eggs, polysaccharides such as arabinogalactan (mustard gum) from mustard seeds, and the protein PLIN4 from milk. Each possesses some amphiphilic character, and each can stabilize emulsions.

Lecithin is an emulsifier in the yolks of eggs, and its ability to stabilize oil dispersed in water is why we see the transformation from a thick liquid to a soft solid in mayonnaise. Figure 5.4a shows a line drawing and a surface charge image of lecithin. We can see from the line drawing that lecithin has a glycerol backbone like triacylglycerides, but the end carbon of glycerol connects to a polar charged group through its oxygen. The group contains a negatively charged phosphate attached to a positively charged nitrogen atom. All this charge makes this part of the molecule very hydrophilic. Meanwhile, the end carbons of the glycerol connect through their oxygens to one saturated fatty acid chain with sixteen carbons and a second monounsaturated fatty acid chain with eighteen carbons. These two fatty acid chains confer a hydrophobic character to the molecule. The surface charge image clearly shows the polar charged head group shaded in red and blue on the left and the nonpolar tails

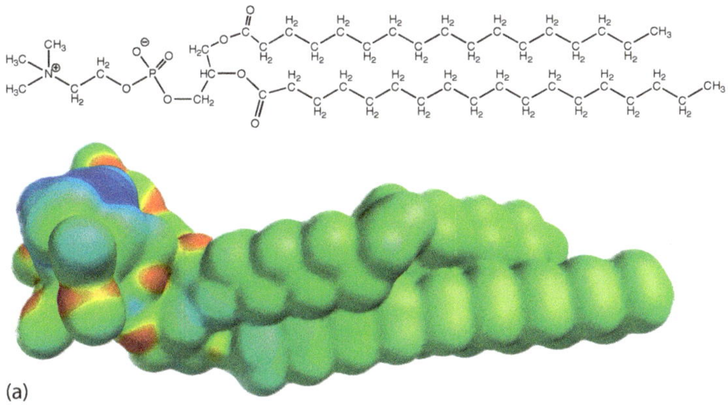

(a)

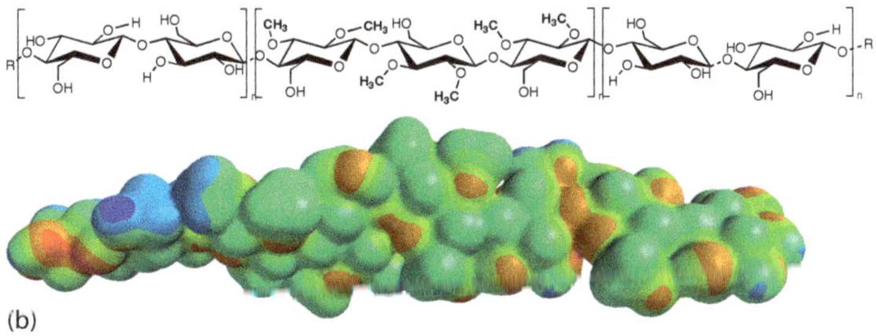

(b)

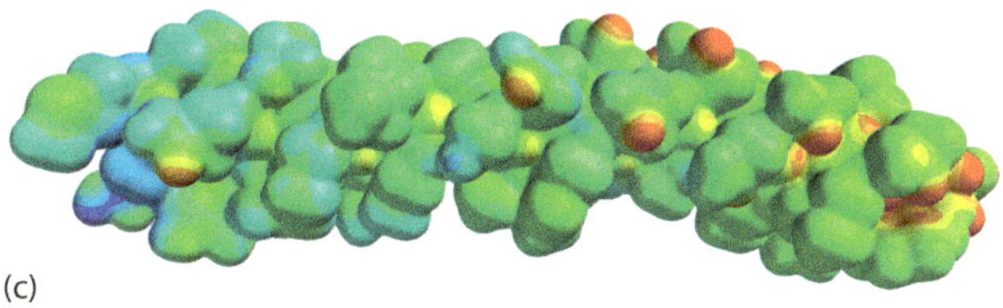

(c)

Figure 5.4 Fats, Carbohydrates, and Proteins Can Act as Emulsifiers

a. Lecithin from egg yolk is a derivatized fatty acid that acts as an excellent emulsifier. It is the lecithin in egg yolks that helps mayonnaise retain its elastic, soft solid state. It is also found in foods such as soybeans and wheat germ. Lecithin is shown as both a line drawing and a surface charge model. It acts as an emulsifier by solubilizing nonpolar substances through its nonpolar fatty acid groups (green surface), while the charged head group (red and blue surfaces) makes favorable interactions with water, keeping the entire mixture stable.

b. A derivatized polysaccharide from mustard seed is a fantastic emulsifier. This is why a teaspoon of mustard is added to most vinaigrettes to stabilize them. Here mustard is shown as both a line drawing and a surface charge model. It has the chemical name arabinogalactan or mucilage gum. The polar and nonpolar surfaces of this emulsifier are not as clearly separated as in lecithin, but it is still possible to see the extended surfaces created by the nonpolar -OCH_3 groups (green surfaces) that make favorable interactions with nonpolar substances and the polar -OH groups (red and blue surfaces) that interact with the aqueous phase.

c. A protein can also act as an emulsifier. Here is a peptide derived from the protein PLIN4, which has 33 amino acids and has the interesting job of lining the surface of a fat droplet and forming an interface between the fat and the aqueous fluids in the blood. As the large size of the peptide (> 200 atoms) makes the line drawing too complicated, only the surface charge image is shown. Amino acid side groups along the helix cluster into nonpolar regions (green surfaces) and polar regions (red and blue surfaces). This pattern facilitates the interaction of the helix with the surface of a fat droplet through its nonpolar side and the aqueous serum through its polar side groups. This type of helix is called amphiphilic.

shaded in green on the right. Together, the hydrophilic head and the hydrophobic tail cause lecithin to situate on the interface between oil and water, stabilizing the emulsion.

Mustard contains a material known as mustard gum, which is isolated from the bran of the yellow mustard seed. The polysaccharide arabinogalactan derived from that gum is responsible for mustard's emulsifier properties. Adding one teaspoon of mustard and shaking the vinegar oil suspension allows the mustard gum to surround the oil droplets and keep them dispersed in the salad dressing. Figure 5.4b shows a portion of the complex molecule that is especially helpful for emulsification. Here, we can recognize glucose molecules polymerized using the beta-type linkage discussed in Chapter 1 to form long polymers like cellulose. One key difference is that some polar hydroxyl (-OH) groups have been replaced by nonpolar methoxy groups ($-O-CH_3$). This modification makes one part of the polysaccharide nonpolar while the other part retains the polarity of the unmodified glucose. Having polar, hydrophilic surfaces alternate with nonpolar, hydrophobic surfaces makes mustard a perfect emulsifier, as it can bind to the oil through its nonpolar surface and to the vinegar/water through its polar surface.

Even the amino acid polymers we know as proteins can act as emulsifiers. The PAT family of proteins, periphilin, adipophilin, and TIP-47, act as emulsifiers essential for maintaining the dispersion of fat droplets in aqueous mixtures. Milk is a suspension of fat droplets in an aqueous solution full of casein and other milk proteins. The fat droplets in raw milk will eventually separate, but they are kept temporarily in solution by associating with the PAT proteins. The N terminal PAT domain consists of alpha-helical repeats of an eleven amino acid sequence that is amphiphilic.[5] Figure 5.4c shows the cartoon structure of a periphilin known as PLIN4. We can see the amphiphilic helix. The side facing us has mostly hydrophobic amino acid side groups shown using the green surface charge shading. This side of the helix is soluble in the oil phase. The other side of the helix has hydrophilic amino acid side groups. We catch a glimpse of the red and blue surface charge shading on the right side of the helix, which suggests that the back side will be soluble in the water phase. The protein coats the surface of the lipid droplet, keeping it in solution. In addition to acting as emulsifiers, the PAT proteins help regulate the complex lipid balance in a fat droplet by substituting for part of the membrane when there is a shortage of phospholipids.

How Do Stabilizers Keep Emulsions from Separating?

Making the continuous phase thicker or more viscous helps keep emulsions stable. Viscosity arises from the internal friction of the molecules against each other and is another important property of liquids that gives rise to different

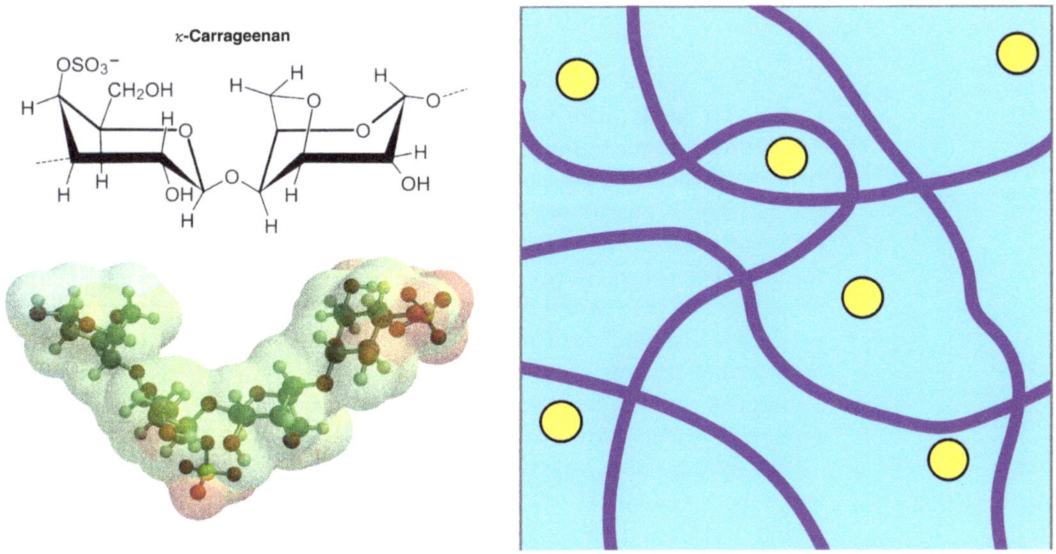

Figure 5.5 How Molecules Like Carrageenan Stabilize Colloidal Mixtures

Carrageenan extracted from seaweed is shown using a line drawing and then as a cartoon. Polymers set up distinct microenvironments in the continuous phase of a colloid (here shown as the blue water background) and prevent the coalescence of the dispersed phase (here shown as yellow droplets of fat.) These stabilized microenvironments are called molecular corrals.

mouthfeels. It is measured as the force necessary to displace a unit volume of the liquid. Certain fluids, such as water, have a low viscosity, and others, such as honey and maple syrup, have a high viscosity. You can increase the viscosity by adding thickeners to an emulsion. These materials include pectin, starch, carrageenan, agar, gelatin, and xanthan gum. When the continuous phase becomes more viscous, there is more friction as droplets move through. The large droplets are broken down into smaller droplets, making them more easily dispersed and preventing them from joining together. As shown in Figure 5.5, carrageenan, the polysaccharide from seaweed, sets up long filamentous strands in the solution that increase the viscosity. The stabilizers also set up a matrix in the continuous phase, corralling particles in the dispersed phases, further preventing them from coming together.

Each stabilizer has unique properties that you can choose depending on the details of the dish. For example, xanthan gum, a stabilizer isolated from bacteria, is stable in both hot and cold solutions and at any pH, and only a tiny amount of it is necessary for thickening. Gelatin is a common stabilizer used to thicken desserts and sauces. Since it is derived from animal products, it is not used in dishes for vegetarians, vegans, and those with cultural or religious restrictions on the consumption of animal products.

SECTION 3: KITCHEN COLLOIDS

You can create delicious foams in the kitchen by dispersing air into a liquid. Forcing air to mix into a solution transforms cream into whipped cream and egg whites into meringue. To make these foams, one must vigorously mix the solution by beating, whipping, whisking, stirring, or churning until it transforms into an elastic solid. Mechanical energy—from the chef's whisking action or a mixer, blender, or foamer—is added to create the foam. In each case, we use mechanical energy to increase the air's volume fraction and shear the micellar droplets created by fats, proteins, or carbohydrates of the liquid. The product is ultimately unstable—and the foam will eventually collapse—unless the mixture transitions to a solid such as butter or a cooked meringue. In this section, we will highlight the background and properties of these colloidal liquids and the transformations they undergo as they turn into foams and more.

The first use of culinary foams dates to the 1700s with the creation of sweet and savory egg-based soufflés. The word *soufflé* is translated as "puffed up," which describes the dish and the soft matter that is neither flowing nor completely solid. The use of foams evolved into meringues and, eventually, the creamers in many gourmet beverages today.

Milk into Whipped Creams

Milk, our first comfort food, is sometimes called the mother of all colloids. It is the traditional starting place for creating the gas-liquid foam we call whipped cream. Milk is mostly water but includes fat, proteins, lactose sugars, and minerals. Milk fats and some proteins like casein do not dissolve in the water but form droplets dispersed in the milk that are large enough to scatter light, making the milk appear opaque and giving it a creamy mouthfeel. To make whipped cream, the starting milk product must have a fat content greater than 35%.

In most of the Western world, adult humans consume milk and milk products from various mammals. This milk can be provided directly to the consumer as raw or processed. The homogenization process forces raw milk through a small opening at high pressure, shearing the large droplets into smaller droplets that resist aggregation. The milk stays mixed and homogenous. Pasteurization, the heating of milk to 72°C for 15 seconds, and ultra-pasteurization, heating to 138°C for at least 2 seconds, are processes for ensuring the long-term safety of milk products away from the farm. The "Drink Milk" marketing campaign encouraged milk consumption, especially by children. Drinking milk is good advice since milk is anticarcinogenic, immunomodulatory, antimicrobial, antihypertensive, helps prevent tooth decay, and lowers cholesterol.[6]

Raw milk is not homogenized or pasteurized. Concerns about foodborne illnesses transmitted by milk, such as salmonella, *E. coli*, and listeria, have

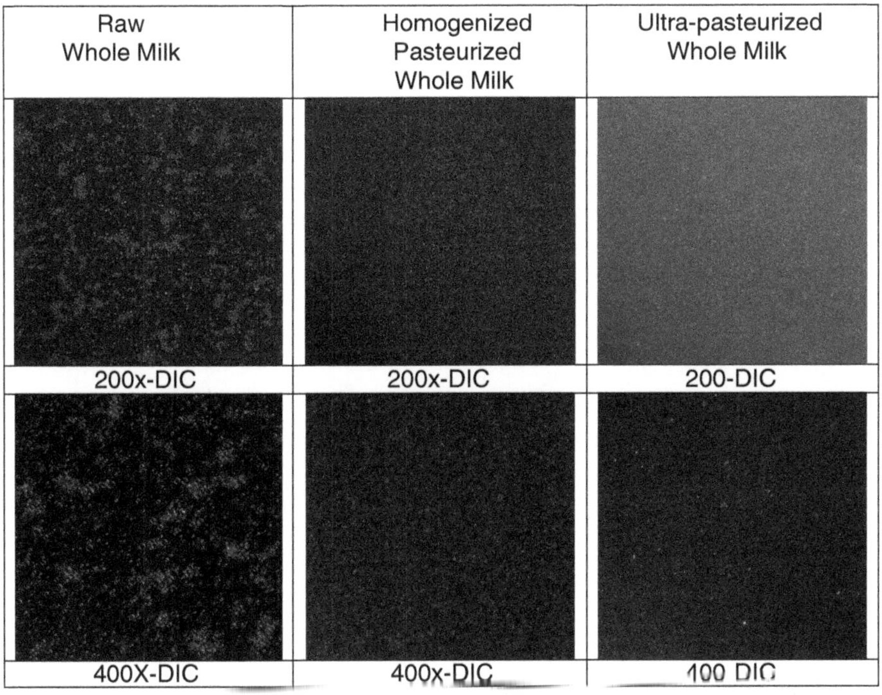

Figure 5.6 Microscopic Imaging of Fat Droplets in Milk

Cow's milk that has not been processed (raw milk), pasteurized homogenized whole milk, and ultra-pasteurized homogenized whole milk are seen at different magnifications by differential interference contrast (DIC). Comparing the raw and processed milk, homogenization reduces the size of the fat droplets. Pasteurization eliminates potential pathogens in the milk. Together the two processes result in the stabilization of the milk colloid but drastic changes in the size and distribution of fat droplets. Even though all three milk products contain 3.5% fat, the size of the droplets changes the mouthfeel and texture.

resulted in the FDA's recommendation against consuming raw milk.[7] Still, some consumers buy raw milk from trusted dairy farms or markets, which some states allow to sell raw milk under very controlled conditions. Raw milk fans describe the product's taste and mouthfeel as unsurpassed. They say that raw milk is better tolerated by those with lactose intolerance, more helpful to those with asthma and osteoporosis, provides beneficial bacteria, and helps build immune systems, especially in children. The FDA challenges these claims.[8]

Raw milk's colloidal composition and large fat droplet size mean it is unstable. Over time, the milk fats and fat-soluble proteins will separate from the watery milk solution, and the cream will rise to the top of the mixture. Figure 5.6 shows the decrease in the size of fat droplets from raw milk to pasteurized, homogenized, and ultra-pasteurized milk.

Casein is a complex protein in milk that comes together to form an encapsulated structure known as a micelle. When treated with acid or enzymes, the casein in milk can form fluffy solids called curds. The protein consists of three isotypes labeled α, β, and κ. They all contain a high percentage of the hydrophobic amino acid proline, which is (in)famous for its ability to disrupt a protein's secondary structure due to the $-CH_2-CH_2-$ side chain that bends back and connects to the N atom displacing an H atom that would have been part of a stabilizing H-bonding network in protein helices and sheets. The side group also restricts rotation about the alpha carbon, making this part of the protein unstructured, as shown by the tan lines in the interior of the protein model.

Serine amino acids with added phosphate groups can modify the κ casein. As negatively charged ions, the phosphoserines are very hydrophilic and extend to the surface to contact water or calcium ions. Calcium ions and phosphate ions form tiny mineral nanoclusters. The remaining portions of the casein chains are hydrophobic and stick together. The light tan strands that make up the interior of the circular micelle in Figure 5.7 represent these hydrophobic unstructured protein regions. A few familiar properties of milk arise from the micellar structure of casein. The large micelles and larger fat globules scatter light, making milk opaque and white. As an aside, the enzyme rennet (from neonatal mammals, microorganisms, or some plants such as thistles) will clip off the extended charged regions of κ casein, making the trimmed micelles less soluble and causing them to aggregate into curds. Adding rennet enables the production of cheeses with unique flavors and textural properties and explains why rennet is the enzyme used most by dairy manufacturers in the United States.[9]

Alternatively, acid can denature the casein proteins, affecting the same curd formation. The soluble milk proteins that make up the whey—the liquid left over after curd formation—are lactoglobulin, lactalbumin, albumin, immunoglobulin, lactoferrin, and lactoperoxidase. This rich protein mix provides extra nutrition when dried and added to smoothies, soups, or bread dough. Whey is a primary source of dairy processing waste and is discussed more thoroughly in Chapter 6.

If you haven't already done so in Chapter 4, check out **Experiment 3: Transforming Milk into Cheese and Yogurt** in culinary explorations in Part IV.

Milk composition varies by species.

Since humans consume milk from many different mammals, comparing the sweetness and fat content of milk from various species is interesting and essential in making culinary or nutritional substitutions or creating new dishes in the kitchen. Table 5–2 shows that at 7% lactose, human milk is one of the sweetest kinds of milk, though it has a low protein content. Cow's milk is slightly less sweet but has three times the protein. Human milk is about average in fat

Figure 5.7 Model of Colloidal Milk

The image shows the molecular complexity that exists in milk. The large yellow droplet extending off the top left edge represents a fat droplet with surface proteins embedded in the outer membrane. Assemblies of the milk protein casein, created from hundreds of proline-rich α and β casein molecules, are represented by the tan spaghetti-like tangles that show the disordered segments on the inside. Negatively charged phosphoserine side groups on the **κ**-casein complex with Ca^{+2} and line the surface of the assembly. The dark brown background represents the protein-rich aqueous continuous phase of the colloid.

Table 5–2 **Variations of Fat, Protein, and Lactose in Milk from Various Mammals**

Species	Fat %	Protein %	Lactose %
Cow	**3.5**	**3.5**	**4.9**
Human	4.2	1.1	7.0
Water buffalo	9.0	4.1	4.8
Goat	3.8	2.9	4.7
Donkey	0.6	1.9	6.1
Elephant	5.0	4.0	5.3
Monkey, rhesus	4.0	1.6	7.0
Mouse	13.1	9.0	3.0
Whale	42.3	10.9	1.3
Seal	49.4	10.2	0.1
Sheep	6.0	5.4	5.1

SOURCE: https://mlaiskonis.wordpress.com/2014/06/01/butter/.

content—which gives rise to its whiteness and mouthfeel. By contrast, milk from the water buffalo has more than twice the fat content of cow's milk, which explains the luxurious taste and mouthfeel of buffalo mozzarella. Two aquatic mammals, whales and seals, produce milk with nearly ten times the fat content of human milk—presumably because they need to provide fat for insulation against cold and for energy storage.

While there are over 8,000 breeds of cows worldwide, four dominate the dairy industry: the iconic black and white Holstein, the more petite light brown Jerseys, the solid gray-brown Brown Swiss, and the brown and white Guernsey. Holsteins lead the herd as milk producers, with each female adult capable of producing 22,000 pounds of milk annually. Milk from Jerseys has the highest protein and fat components—making it ideal for cheese. Milk from a Guernsey cow is golden due to its diet's richness in beta-carotene. Brown Swiss cows are hardy and resilient and are the largest animals of this group—which, with their high milk productivity and rich protein and fat content, make them an excellent dairy breed.[10]

Transforming milk to whipped cream requires high fat.

The standard milk in the refrigerated dairy section of our markets is typically obtained from high-producing cows such as Herefords and is pasteurized and homogenized. Consumers can choose fat content for cow's milk that varies from about 0.5% for skim milk to 4% for whole milk. The dairy shelf also offers half and half (14% fat), light cream (24% fat), whipping cream (30% fat), and heavy cream (approximately 40% fat).[11]

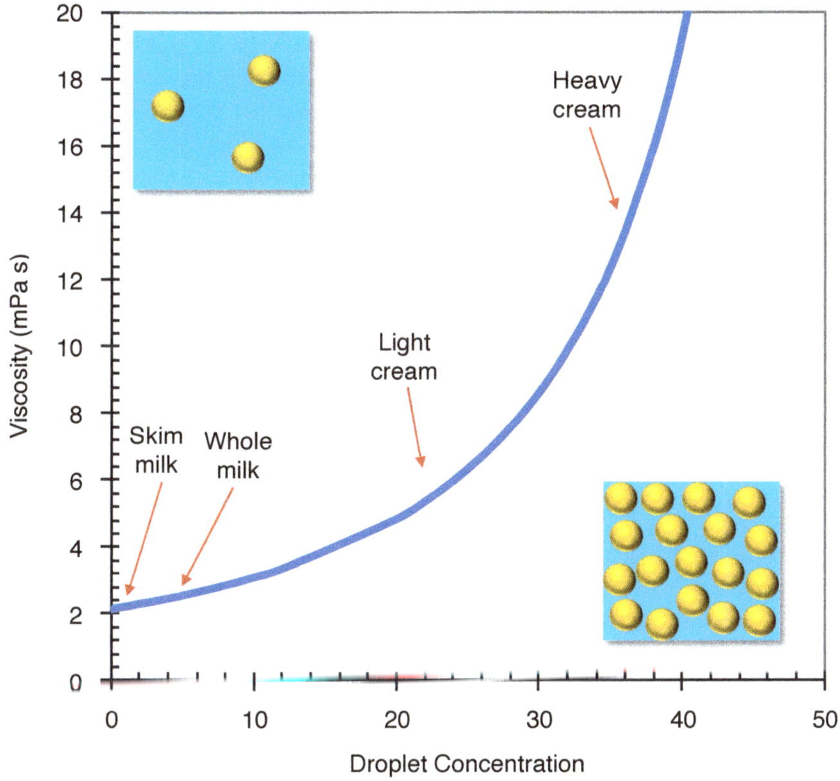

Figure 5.8 Viscosity (Thickness) of Milk Depends on Fat Content

Viscosity or thickness, measured in units of millipascal second (mPa s), is the resistance of a liquid to flow. As the concentration of fat droplets in milk increases from 0.5% (skim milk) to 36% (heavy cream), the viscosity increases exponentially from 2 mPa s to 14 mPa s.

The concentration of fat droplets is one measure of the fat content. As shown in Figure 5.8, increasing fat droplet concentration causes the viscosity to increase, with heavy cream having ten times the thickness of skim milk. Each has a different mouthfeel. As the fat content increases, the concentration and size of fat droplets also increase, and the mixture appears whiter as it scatters more light and has an increasingly creamy mouthfeel. Since transforming milk into whipped cream requires at least 30% fat, chefs must start with either whipping or heavy cream; light cream will not do.

When whipping cream, it is important to chill the bowl and the cream beforehand since the solubility of a gas in a liquid typically increases at low temperatures. While beating the mixture, the air is pulled in and suspended by the sheared fat droplets and protein micelles, specifically casein, surrounding it. The air itself can be considered nonpolar and draws the fat and hydrophobic portions of the amphiphilic protein micelles. The mixture transforms into an elastic solid at a particular volume fraction of air, and soft peaks form. Continued mixing creates stiff peaks with more elasticity. It is essential to know when to stop mixing. If the air volume fraction gets too high, the mixture will break and eventually turn into butter. A liquid solution called buttermilk accumulates on the top surface. If this occurs, note that the process is irreversible. Starting over is the only option.

If you are looking for a quick, time-saving alternative and flavor is not your prime concern, you have several options. The best-selling imitation whipped cream in the United States is Cool Whip, a frozen whipped topping that one must thaw before use.[12] It contains a mixture of vegetable fats, high fructose corn syrup, casein, skim milk, stabilizers, and other food additives that attempt to simulate the flavor and texture of natural whipped cream. Alternatively, you can reach into your refrigerator and grab a can of Reddi-wip. Aaron "Bunny" Lapin invented Reddi-wip in 1948. In nondairy varieties of Reddi-wip, vegetable oil substitutes for the cream and a long list of synthetic additives (polysorbate 60, sorbitan monostearate, sodium stearoyl-2, lactylate, xanthan gum, and lecithin) provides for flavor, texture, and stability. The foam is created as it is dispensed from the pressurized mixture of cream and nitrous oxide, a colorless, odorless, nontoxic gas (also known as laughing gas). As the mixture expands through the nozzle, the cream is whipped into the foam by the expanding gas and can be used to top your pies and sundaes.

Butter itself can be the end goal in processes such as churning. Our ancestors churned butter in a laborious process that involved repeatedly raising and lowering a wooden pestle through a solution of cream until the cream solidified. Some water extruded from the cream can be collected as buttermilk and used in baking. The remaining water is trapped in the solid butter and can be released upon heating, causing the bubbling and frothing you see when you melt butter. Clarified butter has had the excess moisture and some proteins removed, and when used in heated sauces, frothing and browning are eliminated. Ghee is a type of clarified butter used in Southeast Asian cooking that can tolerate high temperatures without decomposing as the milk solids (protein and water) have been removed.

Nondairy Plant-Based Milk

Mammalian milk contains lactose, a disaccharide that requires a specific enzyme to digest. Suppose you are one of the 65% of human adults who are lactose intolerant. In that case, your body no longer produces that enzyme, and drinking milk or consuming dairy products can give you a stomachache. You

Table 5–3 **Nutritional Content of Cow's Milk and Plant-Based Drinks**

Source	Energy	Protein	Fat	Carbohydrate	Fiber	Minerals					Vitamins				
						Ca	Fe	Mg	K	Na	C	riboflavin	B-12	A	D
per 250 ml serving	*kcal*	*g*	*g*	*g*	*g*	*mg*	*mg*	*mg*	*mg*	*mg*	*mg*	*mg*	*µg*	*IU*	*IU*
Cow	67	3.3	3.3	5.4	0.0	125	0	–	–	52	0.5	–	–	208	42
Soy	42	2.9	1.7	3.3	0.4	125	0.45	–	125	42	0	0.21	1.25	208	50
Almond	82	3.5	2.3	13	3.5	175	0.21	71	–	106	0	–	1.06	0	71
Coconut	80	0.67	2.7	14	1.3	133	0.48	–	–	20	3.2	–	0.8	0	53
Rice	113	0.67	2.3	22	0.7	283	0.48	26	65	94	0	0.34	1.51	499	101

SOURCE: B. Walther, D. Guggisberg, R. Badertscher, L. Egger, R. Portmann, S. Dubois, M. Haldimann, K. Kopf-Bolanz, P. Rhyn, O. Zoller, R. Veraguth, and S. Rezzi, "Comparison of Nutritional Composition between Plant-Based Drinks and Cow's Milk," *Frontiers in Nutrition* 9 (2022), https://doi.org/10.3389/fnut.2022.988707.

Table 5–4 **Physical Properties of Cow's Milk and Plant-Based Drinks**

Milk	Viscosity millipascal*seconds	Flow Index	Separation Rate (% h)	Whiteness Index
Cow	3.2	1.00	3.9	81.9
Soy (avg)	4.9	0.95	13.9	72.1
Almond (avg)	13.5	0.76	33.9	67.1
Coconut	47.8	0.40	37.4	67.8
Rice	2.8	0.97	42.8	66.6
Cashew	5.6	0.97	27.5	65.6
Hazelnut	24.8	0.67	1.3	56.3
Hemp	25.0	0.73	4.4	68.5
Macadamia	2.2	1.00	54.4	51.7
Oat	6.8	0.89	40.1	60.2
Quinoa	13.2	0.76	32.0	71.4

SOURCE: D. J. McClements, "Development of Next-Generation Nutritionally Fortified Plant-Based Milk Substitutes: Structural Design Principles," *Foods* 9, no. 4 (2020): 421, doi: 10.3390/foods9040421.

may avoid dairy or take the enzyme lactase as a food supplement beforehand. Some people choose not to consume animal products for religious or ethical reasons. In either case, you have many options: soy milk, coconut milk, almond milk, or any of the drinks listed in Table 5–3. The plant-based milk industry has 41% of the global plant-based processed food products, which are worth more than $10 billion annually. Nutritionally, there is a lot to be said for plant-based milk; soy milk, for example, has fewer calories, far less fat and carbohydrates, no cholesterol, and almost as much protein and even some dietary fiber.

Note that using "milk" to refer to these drinks derived from plants is misleading and somewhat controversial. In some countries, legislation prohibits using the term "milk" for plant beverages. In June 2022, the FDA wrote that it "recognizes there's been a rapid growth of plant-based foods, including plant-based milk alternatives. The FDA wants to ensure that industry can continue developing such products while ensuring consumers understand the products' origin and nutritional content." To accomplish these ends and encourage innovation, it will continue to allow labeling these beverages as milk but will recommend the publication of guidelines for consumers.[13]

When food scientists develop these products, they can manipulate many properties of the milk to get the best stability, nutritional value, and consumer satisfaction. Table 5–4 lists several of these properties. For example, tasting panels show that consumers prefer their milk to be white. When comparing cow's milk's whiteness to others, milk from almond and macadamia nuts falls short, but soy milk is slightly less white than cow's milk. In considering the separation rate—a measure of the milk's capacity to stay as an emulsion—soy milk and macadamia nut milk are far more stable than cow's milk. Another parameter, the viscosity—a measure of the thickness of the liquid or resistance to flow—of cow's milk is among the lowest, while coconut milk is almost fifteen times as thick.

A recent review of plant-based milk analogs states, "These products are complex colloidal dispersions that contain various small particles dispersed within an aqueous medium, e.g., oil bodies, fat droplets, protein particles, or plant cell fragments. The nature of these colloidal particles, such as their composition, structure, size, interfacial properties, and interactions, ultimately determine the physical, functional, sensory, and nutritional attributes of plant-based milk substitutes, including their appearance, texture, mouthfeel, taste, stability, and nutrient bioavailability. Food scientists use the knowledge of the colloidal basis of plant-based milk substitutes to enhance the products' quality attributes, functional performance, and healthiness."[14] The next time you add nondairy creamer to your coffee, be sure to appreciate the hours of research that have produced these delicious milk analogs.

Can Nondairy Products Be Used to Create Whipped Cream–Like Foams?

Since we need 30% fat or more to make a foam, producing a whipped cream from any plant-based milks listed in Table 5–4 is not possible. Fortunately, other possibilities exist. With its high fat content, coconut cream transforms easily into a semisolid whipped cream. Other materials, such as aquafaba, the liquid from chickpeas, contain proteins such as albumin, globulin, saponins, and soap-like amphiphilic molecules. When whipped, the protein structures unravel and, together with the saponins, stabilize air pockets within the mixture by encapsulating them with the hydrophobic side toward the air and the hydrophilic side toward the liquid. Just as with egg white foams, a teaspoon of cream of tartar (salt of tartaric acid) added before whipping can help stabilize the aquafaba foam further by keeping the pH low enough to keep the proteins denatured.

Scientists in Denmark recently introduced an imitation whipped cream from a lactic acid bacterium related to the bacteria used to ferment milk to make yogurt.[15] The whipped topping comprises four ingredients: water, lactic acid bacteria, milk proteins, and a thickener. The bacteria are about the same size as the fat globules in milk and act the same way, lining themselves up around the air droplet and acting as surfactants to reduce surface tension and stabilize the air in the foam. Different strains of the bacteria generate softer or stiffer peaks.

Beating up Air-Liquid Egg Foams: Soufflés, Meringues, and More

Eggs are protein-rich powerhouse foods that are an inexpensive nutritional source. Mastering the many ways of cooking egg-based kitchen dishes is a cornerstone of culinary training, as discussed in Chapter 1. In addition to being cooked by poaching, frying, scrambling, or using sous vide technology, eggs are the central ingredient in air-egg foams like soufflés. Separating egg whites and yolks and beating them individually before gently folding them back together in a batter is one method to achieve a delicate and fluffy texture in cakes like sponges or chiffons. In beating eggs, whether whole or separated, the air is pulled into the egg by mechanical force, and oil and protein particles in the egg surround the air droplets, stabilizing them by acting as a surfactant—lining up along the air-liquid interface in response to their amphiphilic properties and preventing the bubbles of air from collapsing by reducing the surface tension. The mechanical force of whipping forces more air into the mixture, increasing the air volume fraction and shearing the air-oil droplets.

Egg yolks were featured earlier in this chapter in the first discussion of the egg-oil emulsion known as mayonnaise, where oil is the dispersed phase. A cooked mixture of egg yolks and milk creates rich custards and desserts. The egg yolk proteins and natural emulsifiers are thickeners in the egg-milk mixture. Too much heat will cause the mixture to curdle, and lumps of denatured protein will spoil the rich, creamy texture of the custard. Variations in the milk:egg ratio will create custards as liquid as a pourable crème anglaise or as thick as a quiche. If cream is used instead of milk, the resulting custard is a mouth-watering crème brûlée or if churned at low temperatures, a delicious ice cream. Alternatively, starch such as flour can be added to the egg-milk mixture, thus creating a crème patisserie to prepare fabulous desserts such as cream puffs, eclairs, or Boston cream pies. Including starch in the sweetened flour, egg, and milk mix reduces its heat sensitivity, allowing it to boil without causing the egg proteins to curdle. In addition, the starch contributes to the thickening of the mixture as it cools.

Egg whites transform into a soft foam when whisked by hand or with an electric mixer. Forcing air into the egg white increases its volume fraction and decreases the size of the air-protein droplets stabilized by egg white proteins. The foam transitions from a thick translucent liquid to a shiny white colloid with soft and increasingly stiff peaks. The whipping process unfolds the proteins, which then act as emulsifiers in their denatured form. Sometimes adding a weak acid such as cream of tartar or vinegar helps the proteins unfold and stabilize the foam. Baking the whipped egg whites at low temperatures results in meringues and luscious desserts such as pavlova. Some cakes, such as angel food cake, use egg whites alone to mix with the flour and sugar to create, when baked, a light, airy cake often served with fresh fruit or berries.

Stirring up Solid-Liquid Emulsions: The Five Basic Sauces

Any culinary school will surely include training in creating the five basic mother sauces: béchamel, velouté, espagnole, classic tomate, and hollandaise. Each of these is a starting point for making various secondary sauces. These sauces are made by dispersing a solid, such as a roux made of flour and butter or a tomato puree, in a liquid and heating it until the sauce thickens. Additions of other flavoring ingredients allow for taste and variety. The five mother sauces differ from each other based on the different liquids and the different thickening agents. Three of the mother sauces are thickened with roux, a mixture of equal parts of flour and fat. The remaining two mother sauces are thickened with tomato and egg, respectively.

Clarified butter, which does not brown as it is heated, is the chef's choice for making a roux, but any fat will do. The roux is made from equal weights of flour and clarified butter. Commercial butter contains 15% water, so you must lower the weight of flour by an equivalent amount (i.e., to make a roux using 1.0 g commercial butter, use 0.85 g flour). The roux creates a colloidal paste of butter and flour, with the fat serving to separate the flour molecules. As it is gently heated, the flour cooks in the fat and gradually darkens. Cooking the flour in the melted fat for just a few minutes is necessary to let it lose its doughy taste. As you increase the roux's cooking time, it goes from white, used in béchamels, to blonde, used in veloutés, to brown, used in espagnole sauces. Once the color of the roux is right, a hot liquid is added slowly while stirring. The protein and starch molecules in the flour absorb the water from the hot liquid and expand, thus thickening the sauce. Constant stirring while heating prevents lumps of poorly hydrated flour from forming. The roux is done when the sauce coats a spoon. It is important to remember that taking the perfectly cooked sauce off the heat will cause it to thicken as it cools, so it is essential to consider this when serving. No one wants a gelatinous lump of sauce on their dinner dish.

Béchamel sauce, containing only roux (flour and butter) and milk, is probably the simplest of the mother sauces. Hot milk is added to a simple white roux while stirring over low heat until the sauce thickens. The sauce can then be flavored with onion, cloves, and nutmeg and simmered until it is creamy and smooth. Béchamel is delicious in baked pasta recipes like lasagna and casseroles. It is also the basis for some of the most common white, cream, and cheese-based sauces such as crème, mornay, soubise, nantua, cheddar cheese, and mustard.

In making a velouté, the roux is heated while stirred for a bit longer over the heat until it is light brown, giving the roux a nuttier flavor, and blends well with a stock. You may then wish to add herbs, wine, cream, shallots, meat drippings, or a stock made from poultry, veal, or fish. Your Thanksgiving gravy, hopefully without lumps, is a humble member of the velouté family. Other members include Normandy sauce, mushroom gravy, poulette sauce, and herb seafood sauce.

Espagnole is a basic brown sauce, with the roux cooked to a darker brown, bringing a nuttier flavor and a smoky overtone. It is a foundational sauce for making demi-glace, a delicious accompaniment for red meats. A brown stock is added to the roux to make the sauce, and the best brown stock is made by simmering roasted bones in water for several hours. Espagnole also contains tomato puree for a richer flavor and a mixture of minced onions, carrots, and celery known as mirepoix, which deepens the flavor. The key here is the tomato puree, which contains the tomato plant cells that serve as the thickening agent. Sauces such as a red wine reduction, a Madeira sauce, and a port wine sauce are all made from the mother espagnole sauce.

Tomate sauce resembles a traditional tomato sauce but has a more complex taste and a few more steps. You start by rendering salt pork and sautéing aromatic vegetables in the fat. The next step is to add tomatoes, stock, and hambone to the vegetables and simmer them in the oven for several hours. It is typically unnecessary to thicken the sauce with a roux because the tomatoes usually provide enough thickening. Spanish, Creole, Portuguese, and Provençale sauces can be created from this mother sauce.

The last of the foundational mother sauces is hollandaise, a tangy, buttery sauce made by slowly whisking clarified butter into warm egg yolks. As an emulsion of fat and egg yolks, it is a warm cousin of mayonnaise, discussed earlier in the chapter. Since hollandaise sauces use butter rather than oil, it is crucial to use clarified butter to avoid having the emulsion-breaking milk solids contained in butter. Hollandaise is particularly delicious on seafood, vegetables, and eggs. Delicious variations can be made by adding mustard to make a Dijon sauce or shallots, tarragon, and crushed black peppercorns to make a Béarnaise sauce.

SECTION 4: NOVEL EMULSIONS WITH MOLECULAR GASTRONOMY

In molecular gastronomy, the cutting edge of science-based innovative cooking, new textures have been developed through completely new cooking techniques. The Spanish chef Ferran Adrià is credited for much of the change in how emulsions such as foams are used and prepared. In his efforts to enhance food flavor, Adrià eliminated the use of cream or eggs in his foams. Instead, he combined various nontraditional ingredients with air. These culinary foams are often created with unusual flavors from stock, fruit juices, vegetable purees, and even soups. These are combined with stabilizing agents to prevent a breakdown later—stabilizers from natural plant and animal derivatives such as agar, xanthan gum, or gelatin. The result can be a delicious and delicate vegetable or fruit foam with the consistency of cappuccino.

We are familiar with the traditional methods of introducing air into a foam using the mechanical force provided by an appliance such as an immersion blender. We use a specially designed foam whipper called a whipping siphon in the modernist kitchen. This method results in a dense, rich foam comparable to a mousse, and a foam created by this method is often called an "éspuma," or air. A whipping siphon is a sturdy stainless-steel container that is sealed and then pressurized using small gas cartridges. It contains the liquid mixed with a stabilizing agent such as xanthan gum. Pressurized gases are injected from a cartridge into the airtight container. When you are ready to dispense the foam, the container is shaken, inverted, and ejected through a nozzle, like when dispensing a dollop of whipped topping from a Reddi-wip can. As the pressurized mixture leaves the nozzle, the gas expands rapidly, mixing it into the liquid. The stabilizer holds the tiny gas bubbles in the foam, preventing them from collapsing or diffusing. The gas used to charge the siphon is often nitrous oxide, N_2O, or occasionally carbon dioxide, CO_2, which gives liquids an extra fizz. N_2O is harmless, though as a laughing gas, it is also used as an analgesic in minor surgeries. Because of its slightly higher polarity, nitrous oxide dissolves better in the emulsions than air and so can be incorporated at a higher percentage to make even light and airy foams.

As mentioned, a stabilizer is needed to work as a surfactant. The surfactant will coat the molecules of the liquid and reduce the surface tension, allowing the molecules to adhere to each other more easily. Without a surfactant, the walls of the bubble collapse, and the air inside the bubbles diffuses to the surface. The collapse and diffusion create tension, weakening the walls and causing the bubbles to pop and the foam to collapse.

Try it yourself! Check out the instructions for **Experiment 5: Creating Vegetable Foams** in culinary explorations in Part IV.

SPHERIFICATION

Spherification is another modernist culinary innovation that celebrates our delight in textural juxtapositions. A sphere is filled with a single substance and added to a drink or dish. When you chew on the globe, it releases a burst of flavor that enhances the overall experience. One example is bubble tea, in which tapioca encapsulates a fruit flavor that blends with the tea when the globe breaks. Another is chocolate "caviars," in which the chocolate-containing capsule is consumed with ice cream to provide pops of chocolate when consumed with the dessert. Here the capsule is created by the reaction of calcium solution with a solution containing the seaweed-derived polysaccharide carrageenan to form a solid matrix. Small spheres are made by dissolving calcium salts into the food and then dripping that solution into a bath containing carrageenan, which encapsulates the chocolate, fruit concentrate, or olive oil. In reverse spherification, the material to be encapsulated is mixed with carrageenan, and a spoonful of that mixture is submerged in a calcium-containing water bath. This process produces large spheres to offer as palate cleansers between courses or amuse-bouches.

Try it yourself! Check out the instructions for **Experiment 6: Spherification** in culinary explorations in Part IV.

SUMMARY OF CHAPTER 5

In this chapter, we learned how some mixtures of substances are not liquids or solids but colloidal substances or emulsions where one material is dispersed but not solvated in the other. We learned how to describe colloids by their physical properties, such as volume fraction, elasticity, viscosity, and surface tension, as well as their sensory stimulating properties, such as creaminess, mouthfeel, and appearance. We explored the roles of surfactants, emulsifiers, and stabilizers—all of which can be used to stabilize inherently unstable mixtures. We looked in some detail at milk, the mother of all colloids, and eggs and how they can be used to create many different and delicious foods that take advantage of colloidal transformations. We also discussed how food scientists develop plant- or bacterial-based substitutes for dairy products such as milk, yogurt, and whipped toppings. These colloidal transformations are made possible using mechanical energy such as whipping, beating, blending, stirring, and churning. We learned how cooking basic sauces takes advantage of these transformations and how each is prepared and thickened. Finally, we looked at the types of colloids created in the modernist kitchen. We learned how foams and éspumas can be prepared using whipping siphons and compressed gases and how food can be encapsulated in edible balls that pop in your mouth through spherification.

Experiment 5: Creating Vegetable Foams and **Experiment 6: Spherification** in the culinary explorations in Part IV will give you hands-on experience with the ideas and concepts of this chapter.

We are at the end of Part II—Transforming. We have taken a serious look at how food is prepared and have sought to understand this at a basic scientific level. Now we are ready to take a step away from our journey through the techniques used in the kitchen and look toward the future of food. In Part III—Reforming, we will need all the information we have gathered to understand the future challenges of food sustainability in the face of climate change and a growing population.

PART III

REFORMING

In Part III, we look at the visionary talents of scientists and chefs, often in collaboration, who have shown us ways to address future challenges to food production, such as climate change and global warming. Food production must increase in the face of a growing population, and consumption must change to minimize the carbon footprints of agricultural processes.

We first identify some of the causes of climate change that are linked to the food sector and consider the challenges presented by changes in production and yield. We approach this topic using the tools developed in previous chapters. Understanding the fundamental need for fuel and how macronutrients can supply these calories, as described in Chapter 1, helps us understand the carbon balance inherent in the combustion of these fuels, both by living organisms and the machines we have invented.

We consider current efforts to educate consumers and reform eating behaviors that put tremendous stress on the environment and food security. Food preferences, informed by the sensory inputs described in Chapter 2, must guide producers and consumers as more planet-friendly foods are created.

Several research initiatives focus on reducing the burden of waste in our landfills by reclaiming waste streams—a process known as waste valorization. Information presented in Chapter 3 on how to break down the molecules of the primary food groups is helpful here, as is the understanding of how protein enzymes can function at room temperature and in mild conditions to facilitate these reactions in ways that are not harmful to the environment. These strategies will deliver considerable payoffs in the future in reducing greenhouse gas emissions.

We turn to molecular biology to understand how genes can be manipulated to intentionally select traits, such as drought and heat tolerance, shorter growing times, or reduced need for fertilizers, from one variant and move them to another using a new technology known as CRISPR/Cas9. Based on the themes of exquisite molecular recognition and environmental sensitivity we saw in Chapters 2 and 4, we realize these themes play out in a slightly different biological stage in which the players are DNA, RNA, and proteins.

We close by exploring innovations brought to us by food scientists and those trained professionals in the kitchen, ideas that push the boundaries of what it means to be food and to cook food. Not only are plant-based foods reengineered to taste like meat, but hamburgers in the future might be cultured from bovine cells in biological reactors. Not only can delicious entire dishes be produced from digital printers that mix and cook your foods with lasers, but the food itself can be reconstructed from deconstructed food elements. These ideas build on all you have learned about food and its transformations in Parts I and II.

In the end, we want to leave you with the belief that you have the power to effect change and to be hopeful for a future in which many of these climate change challenges will be met.

CHAPTER

6 The Future of Food

Our goal in previous chapters was to understand the basic chemistry of food and its transformations. In this chapter, we step away from this goal and instead attempt to use our acquired knowledge to ask challenging questions about the future of food. We begin by presenting background information in **Section 1: How Climate Change Is Driving Reformation Efforts in Food Systems**. Here we investigate the link between food and global warming. What is food waste, and how does it add to greenhouse gas emissions? In **Section 2: Actions That Can Be Taken Right Now**, we consider a small sampling of ongoing efforts consistent with the Environmental Protection Agency's (EPA's) Food Recovery recommendations. These include educating consumers to reduce surplus food and to make food choices that minimize adding additional stresses to global warming, working with food wholesalers to reduce waste by reclamation of visually imperfect foods and improving packaging and labeling, and changing food donation regulations to save perfectly good food from being dumped into waste streams. Finally, in **Section 3: Promising Initiatives for the Future**, we highlight several new frontiers. We consider the contributions from the molecular world, where scientists have developed new ways of valorizing food waste and preventing spoilage. Then, taking inspiration from microbiology, we show how gene editing using CRISPR/Cas 9 can enhance food production. From those scientists and chefs who imagine entirely new ways of eating, we consider large-scale changes to plant-based diets, the development of cultured meat products, printable edibles, and a new culinary movement known as Note-by-Note Cooking.

This chapter also differs from previous ones in that we do not have all the answers, and in some cases, we do not even know all the questions. We present some of the most relevant challenges—population expansion and climate change—and identify some of the most promising initiatives for addressing these that draw on our understanding of food chemistry and related sciences. We conclude by considering future hopes for providing some solutions to the challenges.

SECTION 1: HOW CLIMATE CHANGE IS DRIVING REFORMATION EFFORTS IN FOOD SYSTEMS

Two indisputable facts are worth considering. First, we will have more people to feed in the future. While it took 123 years for the Earth's human population to grow from 1 billion to 2 billion (1804–1927), it took only 12 years to grow from 7 billion to 8 billion (2010–22). Predictions are that the population will grow to 9 billion by 2037.[1,2] Second, the Earth is warming. Since the industrialization that took place in the late 1800s, the Earth's temperature has risen by 1.1°C, and while this might seem like a small number, the planet's warming is creating changes in the climate and disruptions in the biodiversity that our food production depends on.[3] All climate models predict an even more significant increase over the next few years. Will we be able to provide food to everyone in the face of these disruptions? How, at the same time, will we be able to reduce the impact of producing food on the problem of global warming?

Global warming and climate change happen when the level of certain gases, called greenhouse gases (GHGs), are released into our atmosphere. There they act as a selective filter, letting in sunlight but preventing the release of heat into the vacuum of space. In December 2015, 193 countries and the European Union signed the Paris Agreement, in which they committed to reducing GHG emissions to limit global warming to 1.5°C.[4] Follow-ups to the Paris Meetings include the UN Climate Change Conference of the Parties in Dubai (COP28) in 2023 and in Baku, Azerbaijan (COP29) in 2024. These meetings reaffirmed the original goals while highlighting the global responsibility to provide reparations to the lowest-income countries that, despite contributing in only minuscule ways to the problem, bear the brunt of many of the climate changes, such as rising ocean levels, droughts, and floods. The National Resources Defense Council (NRDC) has written, "The mobilization of support for climate action across the country and around the world provides hope that the Paris Agreement marked a turning point in the global race against climate change. We can all contribute to the cause by seeking opportunities to slash global warming contributions—at the individual, local, and national levels—but we understand better than ever that individual action is not enough."[5]

In this chapter, we will describe the link between food and global warming and then highlight several initiatives that are helping address these problems. We will conclude the chapter by describing several innovations made possible by recent discoveries that hold the promise of having a significant impact on addressing the long-term goal of zero hunger and minimum climate disruption. We will see how scientists, educators, government agencies, and individuals are working together creatively to find solutions to the challenges of food security that are equitable, nutritious, ethical, and planet-friendly.

The tools you have developed in previous chapters will provide you with an appreciation of how to achieve these reforms and, while recognizing the challenges, will convince you that there is reason to be hopeful.

How Is Food Linked to Global Warming?

As can be seen in Figure 6.1, the current sources of GHG emissions to the atmosphere are electricity production (26%—orange), food, agriculture, and land use (24%—green), industry (21%—blue), transportation (14%—teal) and buildings (8%—purple), with all the other sources indexed in gray, totaling 10%. Given its central importance in contributing to global warming and in keeping with our stated goal of reforming the future of food, we will consider solutions to this problem centered in the food and agriculture sector.

How is it that food systems have come to be the second most significant contributors to the problems of global warming? What actions are we taking now to minimize these contributions, and how might the future of food be reimagined to eradicate the imbalance that exists, not only in contributing to the problems of climate change, but also in the food distribution systems that currently leave 10% of the world's population hungry every day. We cannot be complacent. Even though there exists enough food to feed everyone, each year, 2 million children die of malnutrition, and every day, 828 million people, mostly women and children, go to bed hungry.[6]

In 2021, the EPA issued a call to action in its report, "From Farm to Kitchen: The Environmental Impact of Food Waste."[7] It states, "As the United States strives to meet the Paris Agreement targets to limit the increase in global temperature to 1.5 degrees above pre-industrial levels, changes to the food system are essential. Even if fossil fuel emissions were halted, current trends in the food system would prevent the achievement of this goal."[8] Several of the EPA recommendations for change, such as reducing surplus food, changes in food donation regulations, upcycling industrial food waste, and composting food waste at home, will be considered in the following pages. These new ideas will alleviate GHG emissions from incineration and deposition of food waste in landfills.

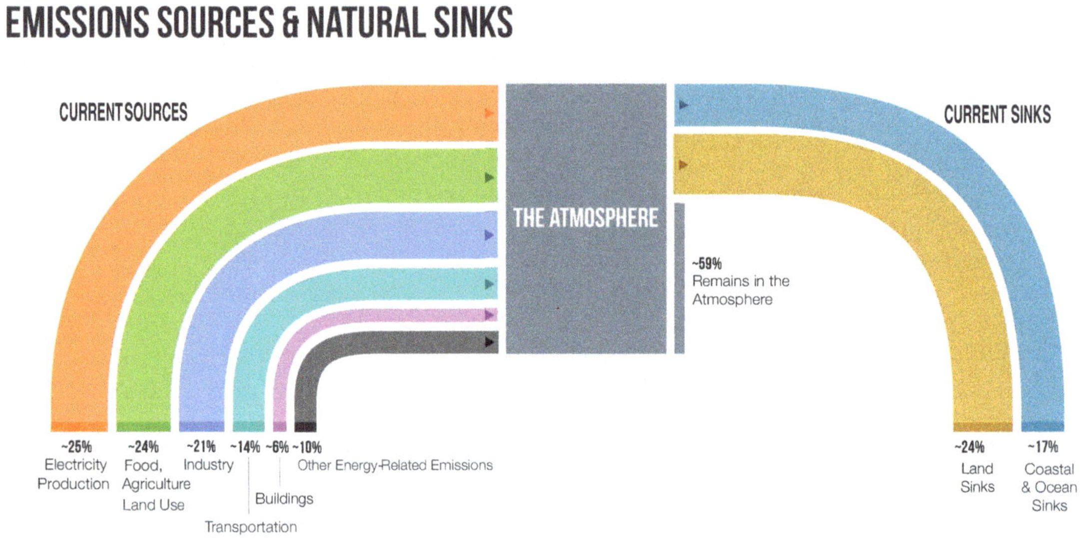

Figure 6.1 Project Drawdown Sources and Sinks

Current sources of greenhouse gas (GHG) emissions are shown on the left as feeding GHG into the atmosphere, while the current sinks that remove GHG from the atmosphere are shown on the right of the diagram. Electricity and the food/agriculture sector produce about half of GHG. Land and coastal areas soak up some but not all of the GHG, leaving about 59% in the atmosphere.

WHAT IS FOOD WASTE, AND HOW DOES IT ADD TO GREENHOUSE GASES?

The United States leads the world in consumer food waste. The numbers are staggering: 40% of the food produced in North America, annually hundreds of millions of metric tons of food, is never consumed but instead moved to landfills as waste. In 2021, the EPA estimated that "each year, U.S. food loss and waste embodies 170 million metric tons of carbon dioxide equivalent (million MTCO2e) GHG emissions (excluding landfill emissions)—equal to the annual CO_2 emissions of 42 coal-fired power plants. This estimate does not include the significant methane emissions from food waste rotting in landfills. EPA data show that food waste is the single most common material landfilled and incinerated in the U.S., comprising 24 and 22 percent of landfilled and combusted municipal solid waste, respectively."[9]

It is sobering to realize that more than 1 billion metric tons of global food waste produce a carbon footprint equivalent to 87% of all road transportation emissions.[10] This waste occurs because of the combined actions of the agricultural, processing, retail, storage, packaging, and retail sectors—with a signifi-

Box 6.1 Food Waste by the Numbers

1.3 billion metric tons: annual global food waste due to production losses and consumer waste
60 million metric tons: U.S. households' annual food waste
47 million metric tons: European Union households' annual food waste.
28%: Share of the globe's agricultural land used to grow food that is wasted.
21%: Share of the U.S. landfill space taken up by food waste.
83%: Share of U.S. households that discard edible food due to misunderstanding expiration date marking.
9.5–12%: Share of European Union households' food waste attributable to expiration date marking.

cant contributor being you, the consumer. How does this contribute to increasing greenhouse gas emissions, and more importantly, how can we all work together to reverse this trend?

Project Drawdown is one agency that has studied the daunting issues related to climate change and specifically focused on what steps will be necessary to draw down the curve of GHG emissions, mitigate the damage done, and propose solutions.[11] It identifies three significant priorities that must be addressed within the Food, Agriculture, and Land Use sector so that future GHG emissions align with the Paris Agreement. Figure 6.2 illustrates these areas: addressing waste and diet by reducing food waste and encouraging plant-based diets, protecting ecosystems by putting in place protection statutes, and shifting agricultural practices by improving efficiency and management while embracing new technologies. According to the estimates from Project Drawdown, over the next three decades, we can contain global warming to a total of 2.0°C by controlling food waste and moving to a plant-based diet and in the process reduce or sequester 88.5 and 78.3 gigatons of CO_2 emissions, respectively.[*,†] In line with the Natural Resources Defense Council (NRDC), Project Drawdown argues that things can change through the combined efforts of policy makers, educators, consumer groups, and innovative products to reduce spoilage and the impact of plastics on landfills. There is room for hope. We will present several initiatives that have already successfully reduced food waste through education, revising our definitions of quality, reducing spoilage, improving food labeling, reducing

* A gigaton is a unit of mass equivalent to one billion metric tons, 2.2 trillion pounds, or 10,000 fully loaded U.S. aircraft carriers.

† A more ambitious goal of holding planet warming to 1.5°C by controlling food waste and switching to a plant-based diet estimates reducing CO_2 emissions by 102 and 103 gigatons, respectively.

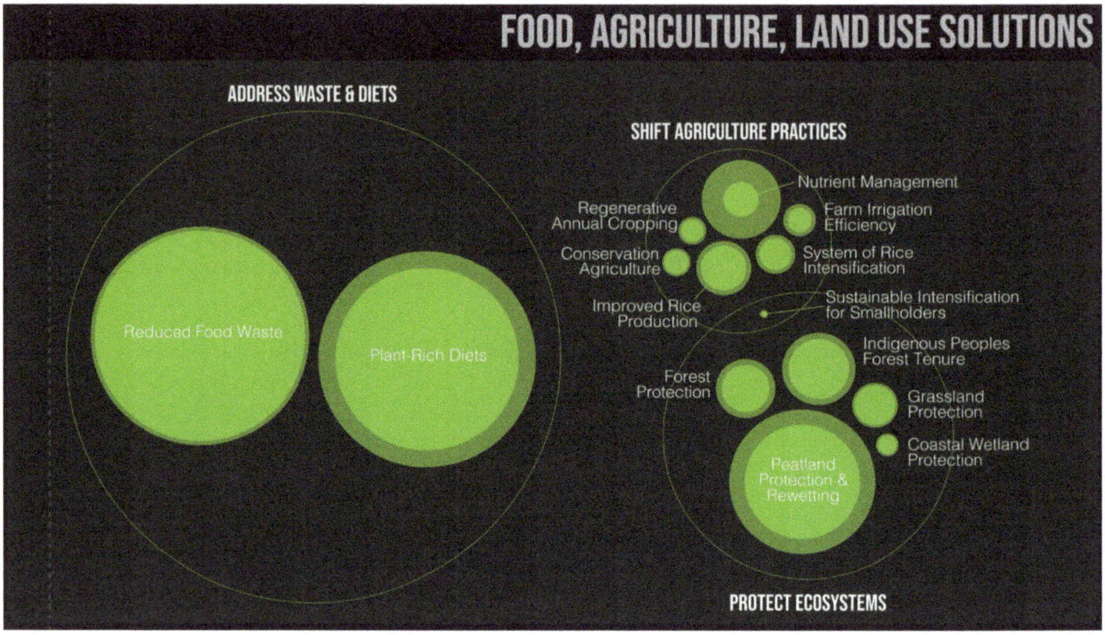

Figure 6.2 Project Drawdown Agricultural Sources and Solutions

Impact of various changes to the food, agriculture, and land use practices on the mitigation of global warming. Reducing food waste and switching to plant-rich diets are seen as the largest factors in reducing GHG emissions. Shifting agricultural practices, most effectively by soil nutrient management and protecting forests, wetlands, grasslands, and peatlands, also contribute. The size of the circles indicates the extent to which this strategy contributes to reducing GHG emissions. The inner and outer green circles correlate with increasing degrees of GHG reduction.

plastic packaging, and reimagining food delivery systems to mitigate food inequities and establish food security. Then we will look into the future to see how scientists imagine the future of food.

SECTION 2: ACTIONS THAT CAN BE TAKEN RIGHT NOW

Solutions exist, and what you do today and tomorrow can make a difference. Below are a few ways we are changing how we interact with food. These examples draw from the education sector (food choices), the private sector (redefining quality, extending shelf life, reducing spoilage, and reducing packaging and plastics), and the regulatory agencies (food labeling and guidelines for food donations). As you read the descriptions of these efforts, recognize how much relies on understanding the science of why we eat, how we make sense of our foods, and how food is transformed by heat, pH, pressure, microbes, and physical means.

Reducing Food Waste Starts with Your Food Choices

Every day, in every meal you eat, you have a chance to play a role in reducing the contributions of the food sector to global warming. In June 2018, one of the premier scientific journals, *Science*, published an article titled, "Reducing Food's Environmental Impacts Through Producers and Consumers."[12] The significance and impact of this article are evidenced by the fact that it has since been downloaded more than 100,000 times and cited more than 2,000 times. The authors make the case that "today, and probably into the future, dietary change can deliver environmental benefits on a scale not achievable by producers. Moving from current diets to those that exclude animal products has transformative potential." The authors go on to highlight that a plant-based diet would have the potential to

- reduce food land use by 76%
- reduce food's GHG emissions by 49%
- reduce acidification by 50%
- reduce eutrophication by 49%[‡]
- reduce freshwater withdrawals by 19%
- remove 8.1 billion metric tons of CO_2 from the atmosphere each year as natural vegetation is reestablished

These numbers are compelling, and while changing to a completely meat-free diet might not be realistic for some, reducing our consumption of meat, especially in the United States, where meat consumption is three times the global average, would go a long way toward reaching these goals. Figure 6.3 breaks down the impact of various food sources and shows the enormous impact of animal meat production, especially beef, on GHG emissions. Recently, newspapers such as the *New York Times*[13,14] and the *Washington Post*[15] have sought to educate consumers regarding their food choices and the impacts of these choices on climate change.

Revising Our Definition of Quality

After inspectors rejected almost 70% of his crop for cosmetic or geometric defects, Mike Yurosek, owner of the California-based Bully-Lov Carrot Farm, was frustrated. It seemed a crime to reject perfectly good produce because of a bump or a bend. In this case, frustration bred inspiration, and Mike asked for a knife and a peeler and went to work "reshaping" the nutritionally perfect but

‡ Eutrophication is the gradual death of an aquatic ecosystem caused by surface algae blooms that result when there is an increase in nutrients such as phosphorus and nitrogen from water run-off, especially from fertilizers and animal waste on farms. The algae blooms reduce the amount of light to the subsurface layers of the water system and ultimately result in the death of all fish and wildlife that the lake supports.

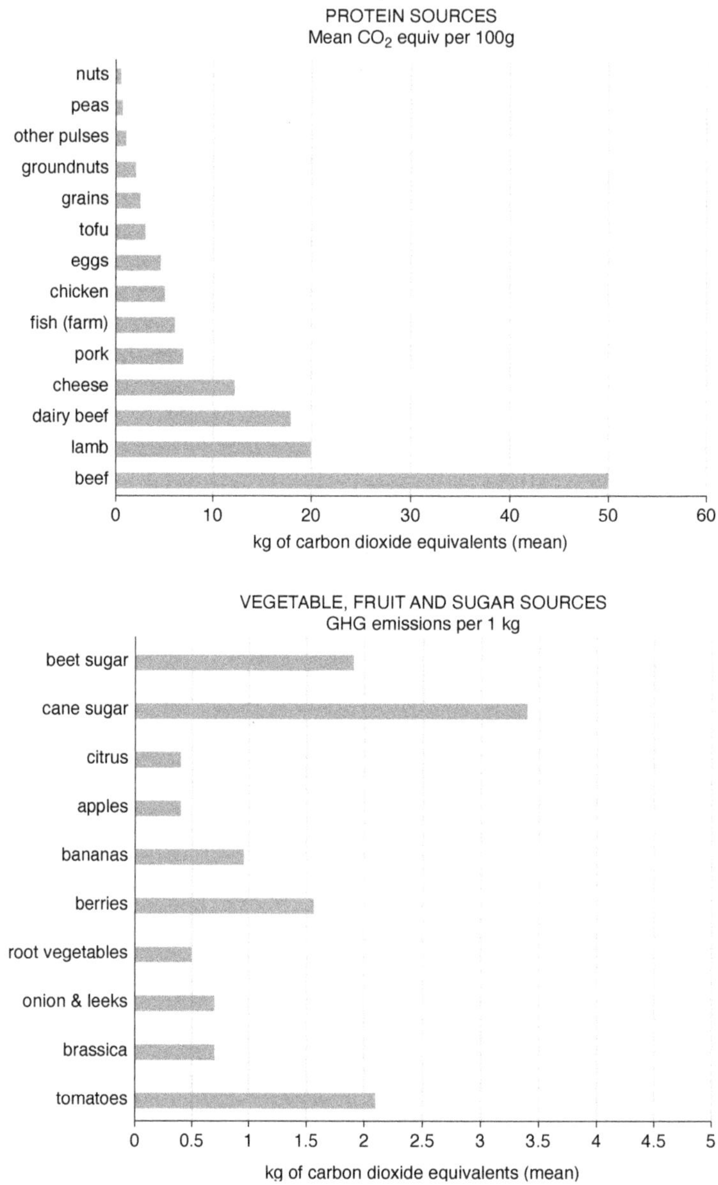

Figure 6.3 Green House Gas Emissions from Various Food Sources

The first bar graph measures GHG emissions from protein sources. Beef production generates 10 times more GHG than chicken production (50 kg vs. 5 kg CO_2 equivalents per 100 g). For dairy and oils, the leading producer, palm oil, produces about seven times more CO_2 equivalents per liter than soy milk (7 kg vs. about 1 kg). Vegetable and sugar production varies in production of CO_2 equivalents per 1,000 g from 0.4 kg (citrus and apples) to 3.4 kg (cane sugar). Starch production shows small variations from cassava (1.4) to maize (0.6 kg). Notice the increasingly small x-axes and the increasingly larger portion.

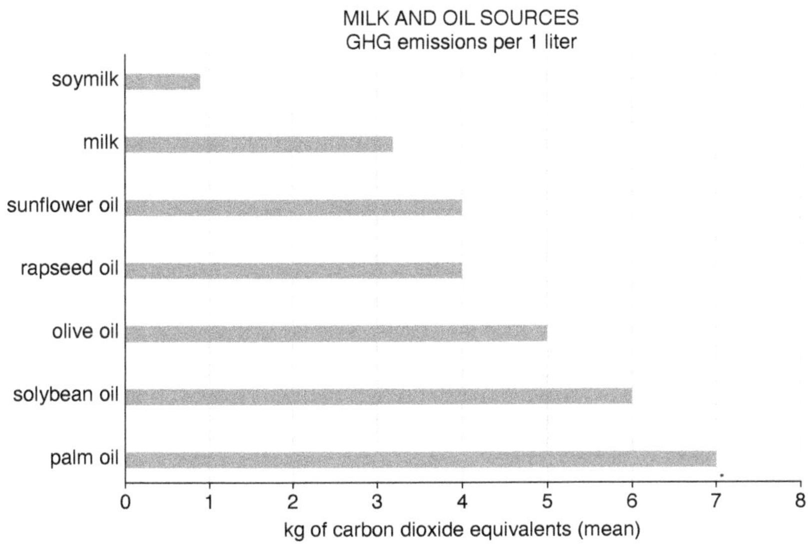

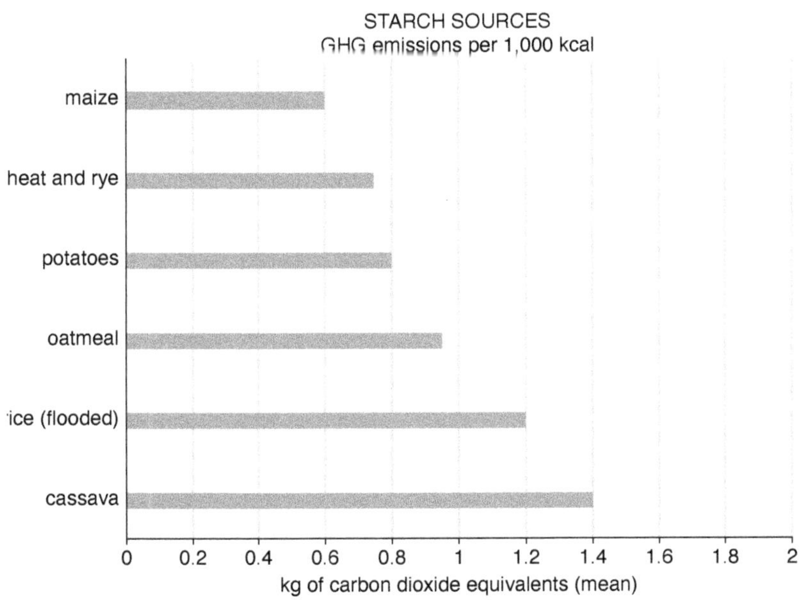

Figure 6.3 *(Continued)*

misshapen carrots into tiny replicas of the fully grown carrots and rebranding them as "baby carrots." That was more than 40 years ago. Today baby carrots make up 70% of the carrot market.[16]

Another company, Misfits Markets, headquartered in New Jersey, is an online grocery service that buys and resells produce that conventional grocery stores reject because it does not meet cosmetic standards or because it is surplus or just past a "best-by" date. Customers can save up to 50% on meat, seafood, pantry items, fruits, and veggies. Starting with four employees in 2018, Misfit raised $200 million from investors in 2021 and by year's end had a valuation of $2 billion. In answer to consumers' question about why they should use this service, the reply is, "Every Misfits Market order benefits farmers and makers, helps prevent food waste, and ultimately helps save our environment."[17]

Three other commercial efforts to bring to market slightly less than perfect but perfectly edible produce are Ugly Fruit (Walmart), I'm Perfect apples, and Spuglies pocked potatoes.[18] A 2016 interview by NPR reported, "Imperfect produce often ends up in landfills instead, contributing to food waste, which, in turn, is a major source of greenhouse gas emissions."[19]

Increasing Shelf Life and Reducing Spoilage

Of course, ensuring the safety of the food we eat is an essential goal. In the United States alone, 1 in 6 people each year develop a foodborne illness, which can sometimes lead to hospitalizations or even death.[20] Labeling foods with "best-by" or "use-by" dates is intended to prevent consumers from eating spoiled or rotten food. To do this, clarity and uniformity in food labeling is crucial. Because consumers are confused about the meaning of these labels, it is common for perfectly good food to be unnecessarily thrown away, thus contributing to food waste. Though no fewer than ten pieces of legislation were proposed in the United States in the 1970s mandating a simplified national system for food labeling, all failed, and instead, states were left to develop their own confusing labels.

The Food Labeling Modernization Act of 2023 (H.R. 2901), was introduced by Sen. R. Blumenthal and Rep. F. Pallone in the 118th Congress.[21] Its stated goal is to amend the Federal Food, Drug, and Cosmetic Act to strengthen requirements related to nutrient information on food labels. The bill includes a broad set of recommendations, including requirements for front-of-packaging (FOP) labels, the format of ingredients, the listing of phosphorus and caffeine content, food allergen and gluten-containing grain declarations, and special regulations for infant and toddler beverages. This important legislation has just started its journey and must pass the House of Representatives and the Senate and be signed by the president before it becomes law. In the interim, we unnecessarily fill our landfills with edible food because of confusing labels. The

United States is certainly not alone in the challenges of producing consistent, accurate food labeling. The European Union's goal to have uniform FOP food labeling by March 2023 is yet to be a reality, as countries have yet to agree on a format.[22]

Meanwhile, companies are developing methods to increase shelf life or delay ripening. The Apeel Company, a relative newcomer to addressing food waste, has developed a coating based on edible mono- and diglycerides that can be economically and safely applied to fresh produce to reduce spoilage. In the company's words, "Our global food system is under severe stress. By 2050, the Earth will be home to a couple billion more people. If we continue using the methods we're using today, we'll need 70% more food and more than 2x the amount of water. So why are we wasting 30–40% of what we grow?"[23]

Fruits such as peaches, tomatoes, avocados, and bananas that continue to ripen after harvest are called climacteric fruit. Other fruits such as olives, watermelon, grapes, and strawberries must ripen on the vine, tree, or plant and are nonclimacteric. For climacteric fruit, two gases, ethylene and carbon dioxide, are naturally and artificially used to regulate the rate of ripening. To delay ripening, producers often keep climacteric fruits cold in evacuated chambers. The development of chemical sensors for the presence of these two ripening gases has helped distributors track the ripeness of their fruit without having to open the evacuated chambers for a visual inspection.

Companies have taken several initiatives to use chemical sensors to monitor the ripeness of food and to clearly indicate to consumers the date by which they should either eat or dispose of the food they have purchased. Determining these transition points in a food's journey from farm to table is difficult. Food chemists must decide which quality and safety attributes are crucial for this food. Microbial growth may be the most important safety concern for some foods, such as salmon or meat. For others, such as olive oil, it might be loss of nutritional value due to oxidation. For bread or pasta, it may be textural changes or mold blooms due to moisture penetrating a packaging barrier. Food quality control labs then measure how the parameter changes over time for a particular food product and calculate a date according to models they build from their data.

Once indices of quality are determined, chemical sensors can be developed. These sensors utilize technologies that monitor gases, contaminants, microbes, oxidized nutrients, and many other hallmarks of aging. Then indicators can be developed that use distinctive visual or tactile signals that clearly inform consumers of the status of produce. Is the product unripe, ripe, or overripe? Does perishable bread, milk, or meat need disposal? Current chemical sensors include light emission or fluorescence to detect degradation or adulteration in foods or package-based solutions, like a sticker that changes color in stages or a milk cap that gets bumpy after exposure to elevated temperatures.[24]

Since we know that ethylene gas controls the ripening of climacteric fruit, is it possible to slow down the process by using another gas to block ripening? Two researchers at North Carolina State University discovered that 1-methylcyclopropene (1-MCP) can slow that ripening process by binding to ethylene receptors in apples. The AgroFresh Company acquired the technology and introduced it first to the apple market and later to six other fruits. Today it claims that using 1-MCP products on over 90% of apples stored in the United States prevents 10 million kg of food waste.[25] Other strategies to reduce the ripening effects of ethylene use ozone to break down the ethylene into harmless product molecules.

One last labeling initiative to highlight is the use of simple stickers or sachets that slow down or stop mold production and reduce food spoilage. A new company, Ryp Labs, has developed "StixFresh" stickers. The CEO, Moody Soliman, claims, "The stickers, which are in the pilot stage, slowly release natural antimicrobial compounds, such as lavender oil, to suppress mold and diseases that spoil food. The company is also developing bigger labels, sachets, and infused packaging materials to deliver the compounds."[26]

Reducing Plastic Packaging

Food packaging, wrappers, straws, cutlery, and bottles, especially those made from polyethylene or polystyrene plastics, are clogging up our landfills and oceans. Biodegradable plastics help a bit, but they still take several years to decay. Newly identified microorganisms such as *Ideonella sakaiensis* that use plastics as a digestible food source may be the key to reducing the environmental burden caused by durable plastics in the future,[27] but what can we do now?

The food giant Nestlé has pledged to reduce its use of virgin plastic wrapping by 30% by 2025 and use only recycled plastic or other renewable resources.[28] But how is the plastics industry changing to accommodate new environmental concerns?

Edible packaging, though hardly a new idea, is one growth area. Foods wrapped in an edible container abound. For example, tacos, hot dogs, and falafels have grain-based containers, and for centuries, sausages have been encased by the linings of animal intestines. Foods like tamales or laulau are baked in inedible but biodegradable packaging (corn husks and tarot leaves, respectively). The edible cone, a baked sugary wafer-like conical container for food, has been with us since the ancient Greeks and Romans. Nineteenth-century French paintings and British cookbooks reference coronet-like holders for desserts. Many believe, however, that ice cream cones went mainstream when they were introduced at the St. Louis World's Fair in 1904.[29] A hundred years later, we are still hoping to expand these strategies and develop edible packaging to significantly reduce the food packaging burden on landfills. Fortunately, several companies are making great strides in doing just that.

Today we can find examples of edible packaging everywhere. The British company Notpla pioneered using seaweed alginate as a palatable plastic replacement in its edible container, Ooho. Construction of these containers uses the same precipitation reactions between sodium alginate and calcium described earlier in this book to make culinary spheres. At mile 23, runners in the 2020 London Marathon received Ooho-packaged lime-flavored sports drinks rather than water-filled plastic cups. In this instance, the edible drink containers or pods saved London streets, lawns, parks, and gutters from being littered with the thousands of plastic cups that would otherwise have been discarded and ended up in landfills. These pods are now used in restaurants throughout London to contain condiments such as ketchup, mustard, and mayonnaise. At a London whiskey-tasting event, attendees were treated to samples of Glenlivet whiskey in Ooho pods. The U.S.-based company Loliware uses similar technology to manufacture straws, and the Indonesian company EvoWare is fashioning burger wrappers out of alginate packaging. Edible cupcake liners can be made with potato starch. The milk protein casein can be reengineered to make a clear food wrap with high strength that is more effective at blocking oxygen than a typical plastic wrap. Fibrous wheat and barley remnants are being refashioned into six-pack loops for packaging beer, and though perhaps not edible, these products degrade in months and can even—in some instances—be a source of food for terrestrial and aquatic animals.[30]

Responding to Challenges to Food Security

One-third of the food in the United States goes uneaten. However, the U.S. Food and Drug Administration only recently modified its Food Code to address food donations specifically.[31] This revision was part of the National Strategy on Hunger, Nutrition, and Health, which provides a road map of actions that the federal government will take to end hunger and reduce diet-related diseases by 2030—all while reducing food disparities.[32] The 2022 Food Code has clarified that food stored, prepared, packaged, displayed, and labeled according to Food Code safety guidelines can be donated. This change helps prevent food loss and waste across the food supply chain and helps ensure safe, good-quality food gets to those who need it most.

One Generation Away is one example of many nonprofit companies that collect and redistribute food. It is a locally focused organization founded in 2013 by Chris and Elaine Whitney that serves the southeastern United States. It collects high-quality food at or near its expiration date and redistributes it to food banks and mobile pantries. From there, food items are delivered directly to those in need. It also provides food to individuals in disaster areas.[33]

A much larger global agency, **World Central Kitchen (WCK)**, also seeks to redistribute food and provide meals, especially to individuals during prolonged humanitarian crises.[34] Michelin-starred chef, José Andrés, has described how in 2005 he watched in frustration the footage of survivors of Hurricane Katrina in New Orleans's Superdome struggling to find water and food. He knew we could do better. When an earthquake struck Haiti in 2010, causing 220,000 deaths and devastating the country's infrastructure, Andrés was ready. Within days, he was on the ground in Haiti, working with local communities to organize and feed their people. His distribution principles are location-specific, regionally sensitive, and empower local leaders to establish food and distribution systems that make sense.

Most recently, WCK has mobilized quickly and strategically to address threats to food security resulting from man-made and natural disasters. One year after the Russian invasion, WCK had provided over 210 million meals in Ukraine.[35] In the month following the horrific February 6, 2023, earthquake in Türkiye and Syria, WCK served over 12 million meals in what it described as its most complex operation, with over 750 distribution sites.[36] More than a million meals have been distributed in the Middle East since the most recent escalation of violence. Since its inception, World Central Kitchen has served hundreds of millions of meals all over the globe. It is first to the front, providing meals in response to humanitarian, climate, and community crises and working to build resilient food systems with locally led solutions. Chef Andrés believes that food is a universal human right, and he has worked to put political pressure on establishing a permanent cabinet position, Secretary of Food.[37]

SECTION 3: PROMISING INITIATIVES FOR THE FUTURE

The solutions cataloged in the previous section developed by clever scientists, farmers, entrepreneurs, businesspeople, government regulators, and chefs have significantly reduced food waste and distributed food more equitably within the current systems and paradigms. What about creating new paradigms of food production? What does the future of cooking look like when we capitalize on new scientific inventions and approaches and bring them to bear on these food security problems in a changing world? Here is a sampling of future-focused solutions from the disciplines of material sciences, bioengineering, and food science.

From the Molecular World: Valorization of Nutrient-Rich Waste Streams

Immobilized Enzymes

Julie Goddard's research in the Department of Science at Cornell University seeks to improve food quality, sustainability, and food safety.[38] Her research publications on active packaging, biocatalytic materials, nonfouling surfaces, antimicrobial coatings, and chelating materials can be found in the *Journal of Polymer Science*, the *Journal of Physical and Colloid Chemistry*, the *Food and Agriculture Journal*, and *Analytical and Bioanalytical Chemistry*. Here we will explore one aspect of her work: how immobilized enzymes or proteins can help with food waste.

In a 2018 review article in the Nature Partner Journal *Science of Food*, Goddard identifies the preconsumer area of food processing waste streams as being particularly open to innovations to reduce the burden of food waste by reclaiming valuable commodities from otherwise useless waste products—a process known as upcycling.[39] Her analysis highlights how producers can use immobilized enzymes to transform waste streams from nutrient-dense foods. It also suggests ways to transform each waste stream into a revenue stream. By looking at how each of these nutrients can be enzymatically transformed into a revenue product, we can see hope for ways to address our challenges of global warming and sustainability.

Waste cooking oils can be upcycled using immobilized enzymes such as lipase, which hydrolyzes the ester bond that we have seen connects a fatty acid to glycerol. By reacting the starting waste oil with alcohol, the enzyme catalyzes a reaction to produce a combustible biodiesel fuel without forming soap, a contaminating by-product when using chemical processes. Reactions like this can operate under green conditions (low temperature and no harsh solvents) or even solvent-free conditions with less harmful environmental impact. The enzyme itself can also be recovered and reused, adding to the economic feasibility of the method.

Dairy waste, such as whey, is rich in carbohydrates and sugars. One of these waste products is the disaccharide sugar lactose. Figure 6.4 shows how a two-step enzymatic transformation can convert this sugar first into the monosaccharide galactose by hydrolyzing the glycosidic bond between the two sugar monomers and then by isomerizing the galactose into the much sweeter tagatose. Both enzymes, β-galactosidase and l-arabinose isomerase, can be immobilized and chemically active in waste streams. Further modification of the sugars by immobilized enzymes can also create surfactants or biodiesel candidates. Higher molecular weight sugar polymers such as starch are found in waste streams from corn, rice, or potato processing. Starches can be hydrolyzed to monosaccharides or acylated to form texture modifiers or plasticizers.

The last major nutrient group to consider is proteins, which are amino acids held together by a peptide bond. Proteins contain dozens to thousands of amino

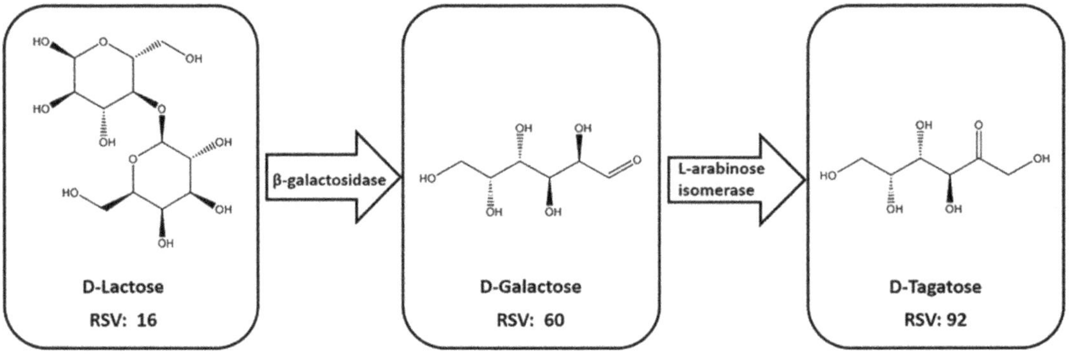

Figure 6.4 Valorization of Sugars from Dairy Waste

A two-step valorization process that uses enzymes to convert lactose from dairy waste, which has a relative sweetness value (RSV) of 16, to tagatose, a monosaccharide with an RSV of 92.

acids. Protein waste comes from many food processing sources and, in the simplest case, can be broken down into smaller amino acids or peptides using enzymes such as the digestive enzyme trypsin. Trypsin catalyzes the breaking of the peptide bond by adding water (hydrolysis). It specifically hydrolyzes a peptide bond between one amino acid and a basic amino acid like arginine and lysine. Whey protein from dairy waste is successfully hydrolyzed by immobilized trypsin. The hydrolyzed protein can provide umami flavors in processed foods, such as soups and sauces, and meat products, such as chilies and gravies.

Radical Scavenging Reduces Spoilage and Mitigates the Need for Food Additives

No one likes spoiled food, not the consumer or the retailer or the wholesaler. Yet health-conscious consumers are demanding the removal of artificial food stabilizers from their foods. In a 2020 article in the journal *Food Control*, Goddard reports on a new method her lab is working on to immobilize natural food stabilizers from protein derivatives on polyethylene plastic wrappers.[40] The two-step process, shown in Figure 6.5, involves a UV-ozone treatment plus an immobilization step, which uses a polymer polyethylenimine to tether a protein to the plastic wrap. Initial studies showed only moderate activity against oxygenated species known as radicals that cause food degradation. Still, it is a first step and shows some promise for the future of wrapping agents that can prevent food spoilage.

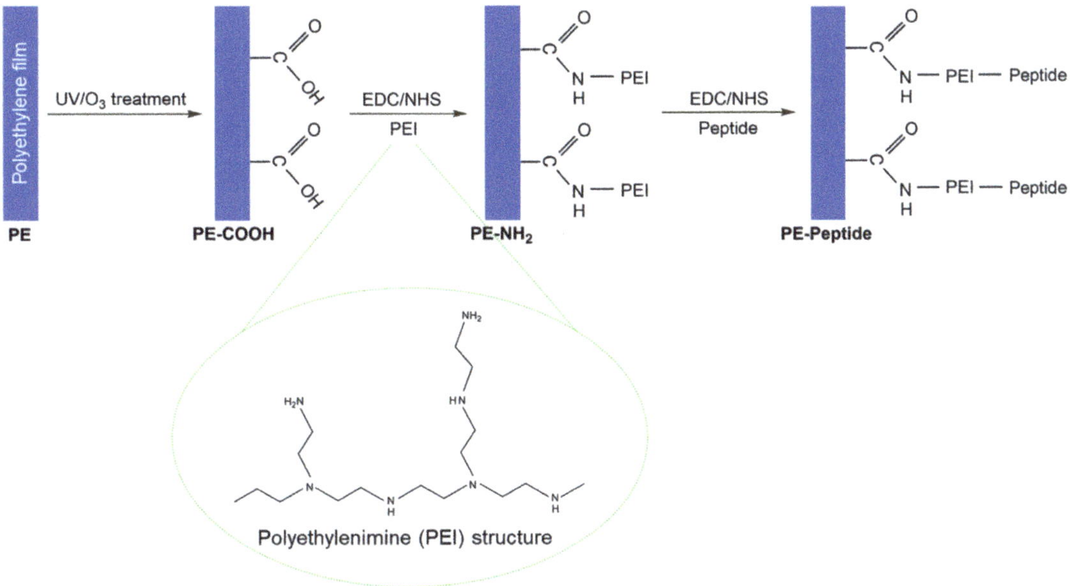

Figure 6.5 Radical Scavenging Food Wrappers

Reactive peptides can be added to polyethylene (PE) using a three-step process: first use UV light and ozone to oxidize the PE to create a carboxylic acid derivative (PE-COOH), then treat this with polyethylene imine (PEI) to join the two to make a amine reactive center ($PE\text{-}NH_2$), and finally, add peptides that form an amide bond to the $PE\text{-}NH_2$ to yield polyethylene peptides (PE-peptide). This activated polymer can be used to make radical scavenging film wrappers that will minimize food spoilage. In the conversion, EDC and NHS are chemicals that facilitate the coupling.

Abbreviations: EDC: 1-Ethyl-3-[3-dimethylami-nopropyl]-carbodiimide hydrochloride; NHS: N-hydroxysuccinimide

Degrading Mixed Plastics Using Chemistry and Microbiology: A One-Two Punch

A real challenge in plastic recycling is that landfill waste is a jumble of different types of plastics. These mixed plastics can include high-density polyethylene (HDPE) used in plastic milk bottles, polyethylene terephthalate (PET) used to package water or soda bottles, and polystyrene (P.S.) used for hot and cold coffee cups. Recently, a group of scientists led by Gregg Beckham, a chemical engineer at the U.S. National Renewable Energy Laboratory (NREL) in Golden, Colorado, devised a strategy to deal successfully with mixed plastics.[41] As shown in Figure 6.6, the mixed waste plastics are first oxidized to organic acids using metal catalysts. The products are then further converted using a process of microbial degradation called bioconversion. Incubation with a genetically engineered soil bacterium, *Pseudomonas putida*, produces two products that are less harmful to the environment. The products are also useful as starting materials for other industrial applications. The versatility of

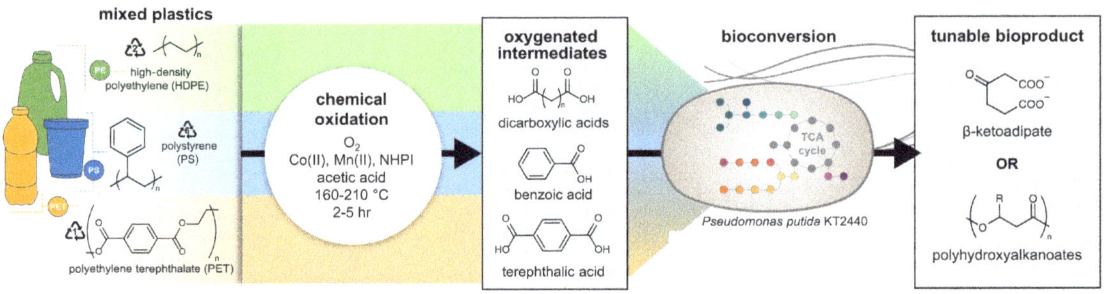

Figure 6.6 Upcycling of Mixed Plastics Using Combined Chemical and Biochemical Conversions

Recovery of useful raw materials from a single-stream recycling mixture uses a tandem chemical and biochemical conversion system. Without sorting, mixed plastics (high density polyethylene [HDPE], polystyrene [PS], and polyethylene terephthalate [PET]), are chemically oxidized by exposure to oxygen under acidic conditions with heating to create mixtures of intermediates. This mixture can be biochemically converted to useful products with a bacterial strain of *Pseudomonas putida*.

this approach will significantly reduce the pressure on landfills from food packaging.

From the World of Microbiology: Gene Editing Using CRISPR/Cas9

Imagine transforming a plant's genetic material to insert traits for disease resistance, drought tolerance, more rapid growth, or higher grain yield! Such transformations have been made possible by a powerful new gene editing technology, CRISPR/Cas9. Agricultural scientists and microbiologists are working to develop food sources that will be more sustainable in the face of climate change and growing populations. The Nobel Committee awarded the 2020 Nobel Prize in Chemistry to the CRISPR/Cas9 pioneers Jennifer Doudna and Emmanuelle Charpentier.[42,43] Nobel Prizes are usually awarded for accomplishments that have stood the test of time. This unusual step of the committee to honor research done within the past decade shows that they recognized the power of CRISPR/Cas9 to change the world. Before describing the details of this new technology, let us review some basic ideas from biology.

The core of molecular biology is the central dogma theory: genetic information flows in only one direction, from DNA to RNA to protein or from RNA directly to protein. DNA (deoxyribonucleic acid) contains a nearly indestructible archive of heritable instructions for how to make an organism. These instructions are coded by the unique sequence of the molecules called nucleotides: adenine (A), guanine (G), cytosine (C), and thymine (T). Since all of these nucleotides share the chemical trait of gaining a hydrogen ion in solution, they are called bases or nucleobases. In DNA, these nucleotides are attached to a sugar-phosphate molecule to create the nucleosides adenosine, guanosine, cytidine, and thymidine. The structure of adenosine triphosphate, or ATP, a related molecule in this family that functions as energy currency is shown in Figure 1.1, in Chapter 1. Here we consider adenosine and other nucleosides' completely different roles as core elements in the genetic code. DNA is a polymer of nucleosides joined together by a linkage known as a phosphodiester bond. Like the twenty amino acids that make up proteins, the four nucleobases are the molecular building blocks of DNA. Each organism has a unique nucleotide sequence, which acts as the blueprint for its construction and functioning. A strand of DNA can contain billions of nucleotides. One strand of DNA will bind to a second strand with a complementary sequence through a very specific set of noncovalent hydrogen bonding interactions to make a DNA duplex. Two strands are complementary when the nucleotides on one strand interact with the other through the specific base pairing rules of A with T and C with G. This complementarity allows for the stable maintenance of the genetic information and the replication of DNA. An organism's entire list of nucleotides is called the DNA sequence or its genome. All 3 billion base pairs of the human genome were sequenced in April 2003. In terms of genome size, that puts humans right between an olive tree (1.4 billion base pairs) and a lungfish (48 billion base pairs).

For an organism to function, it must take this coded information in the genome and use it to make the enzymes and structural material for the cell. Genes are stretches of DNA that contain instructions for how to make a particular protein. The double helix of DNA must be unwound for the DNA to be read. With the help of an enzyme known as a polymerase, the sequence of DNA bases is transcribed into a complementary strand of RNA (ribonucleic acid). RNA differs from DNA by adding a single sugar alcohol and a minor change in one of the bases—replacing a hydrogen for a methyl group that changes a T into a U (uracil). Transcription is highly regulated. As you can imagine, turning genes on and off is crucial to the proper functioning of an organism. Next, the ribosome, a large protein–RNA complex, binds the RNA message and reads it—translating each set of three bases known as a codon into a specific amino acid. In one example, the central dogma is at work as we watch the DNA sequence ACC transcribed into the RNA sequence UGG and translated into the amino acid tryptophan. The messenger RNA is read codon by codon, and the protein is made amino acid by amino acid and finally released when the ribosome encounters one of the STOP codons, UAG, UAA, or UGA.

How does CRISPR technology work? CRISPR, short for "clustered regularly interspaced short palindromic repeat," is a unique piece of RNA designed to bind two things: a DNA-cutting enzyme called Cas 9 and a complementary piece of DNA on the host genome. When the DNA from the host binds to the CRISPR RNA, it is sliced apart at a particular location by the Cas enzyme and released back into the cell. Meanwhile, a new gene can be added to the cell, and natural repair enzymes will insert the new gene into the cleaved host DNA. This new genome-editing technology allows scientists to insert desirable genes,

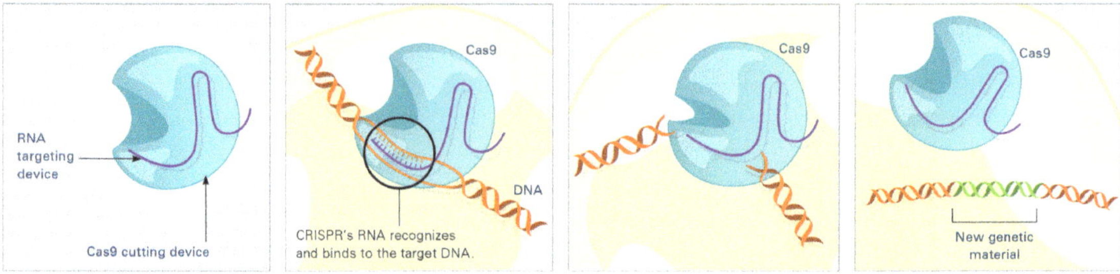

Figure 6.7 Four Steps in CRISPR/Cas9

Existing genes can be edited to insert new desirable genetic material from natural variants. Panel 1 shows the CRISPR/Cas9 protein as a blue sphere with a mouth-like cutout, the active site, with a bound RNA, represented by a purple string. Panel 2 shows a piece of double-stranded DNA, represented by two intertwined orange ribbons, that unwind on binding to the RNA. Panel 3 shows the protein closing on the complex, cutting, and releasing two pieces of DNA. Panel 4 shows green ribbons of double-stranded DNA inserted in the broken DNA to create new genetic material.

such as those for drought or heat tolerance, into a plant host genome quickly, efficiently, and cheaply.[44]

Figure 6.7 details the critical steps in this technology.[45] First, as in Panel 1 of Figure 6.7, an RNA protein complex (CRISPR/Cas9) is constructed that contains a piece of RNA complementary to a specific piece of host DNA and an enzyme to cut DNA. Next, as shown in Panel 2 of Figure 6.7, the RNA protein complex is inserted into the host and binds to the complementary DNA region, forming an even larger DNA-RNA-protein complex. Panel 3 of Figure 6.7 illustrates how the bound DNA is cut into two fragments by the Cas enzyme and released to the cell. Finally, as in Panel 4 of Figure 6.7, a new piece of DNA containing genes for the new trait is made available to the cell and reconnected with the two complementary host DNA fragments, and the natural repair enzymes within the cell repair the broken phosphodiester bonds to make an intact DNA that is available to the host machinery for transcription, translation, and expression.

Unlike transgene products, the DNA is a natural variant within the plant family, so the technology does not exhibit the instability of transgene products. Because these modified plants use DNA from natural variation within the plant family, the CRISPR/Cas9 plants are not considered genetically modified (GM) Consequently, they are not subject to restrictions on those foods except in the EU.[46]

While there is great interest in using CRISPR/Cas9 technology in human health and disease, we focus here on how this technology can help with food sustainability. In 2017, two of the largest companies holding foundational CRISPR patent portfolios, Pioneer-DuPont[47] and the Broad Institute, joined forces to license CRISPR technology. Both agreed to jointly provide nonexclusive licenses for commercial use of the full suite of foundational CRISPR patents in crop agriculture.[48] Several companies have been formed to develop

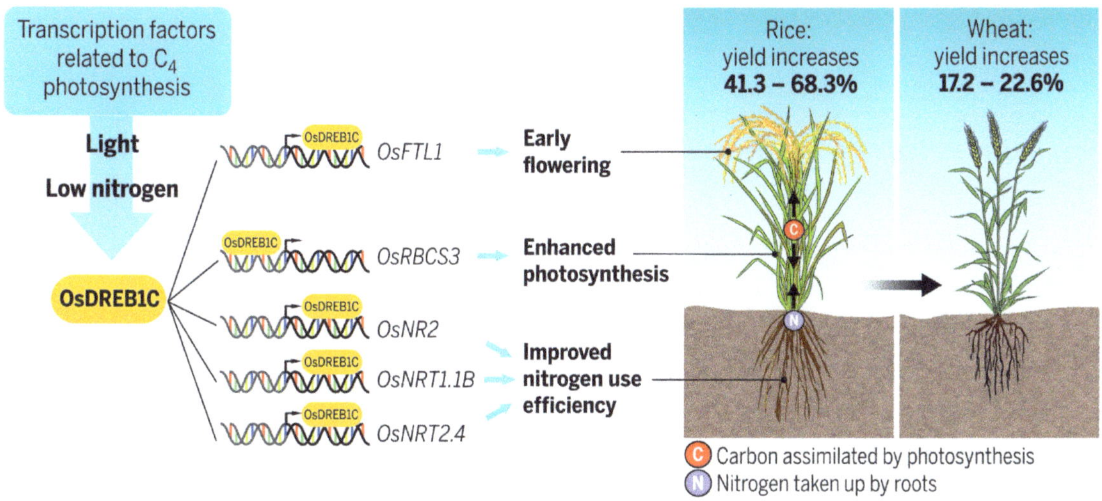

Figure 6.8 CRISPR Genes for Better Plant Growth

Enhanced growth and crop yield in rice and wheat can be accomplished by using CRISPR/Cas9 to increase the expression of the transcription factor gene (OsDREB1C). This transcription factor is activated by light and low nitrogen to stimulate the expression of five target genes (OsFTL1, OsRBCS3, OsNR2, OsNRT1.1B, and OsNRT2.4) in rice and wheat that will cause them to flower early, exhibit enhanced photosynthesis, and improve nitrogen use efficiency.

genetically modified plants that will mature faster, live longer, tolerate heat better, need less water, and hundreds of other traits that would benefit our food supply, especially in the case of climate change. As of April 2022, twenty-seven companies were using CRISPR/Cas9 technology for modifying plants as diverse as cotton, carrots, cucumbers, and cannabis. The U.S.-based companies Inari Agriculture[49] and TreeCo[50] are already developing more sustainable and commercially profitable variants.

These new genetically modified plants can, for instance, insert or modify genes for nitrogen fixation in a plant, thus minimizing the need for fertilizers. Inari focuses on producing seeds requiring less land, water, and fertilizer. They also hope to design seeds that protect themselves against pathogens, mitigating the need for insecticides and fungicides. The system they have developed to adapt seeds to these various challenges is named SEEDesign.

The research literature abounds with crucial genes to consider. Transcription factors are proteins that bind to DNA and act like switches to initiate or accelerate (or turn off) the copying of specific genes. If these genes are related to photosynthesis, this represents a natural target to improve crop yield. Recently, one such transcription factor, with the name OsDREB1C, was found to affect yield and growth duration.[51] As shown in Figure 6.8, when activated by light or low nitrogen, the genes related to photosynthesis are turned on and result in the

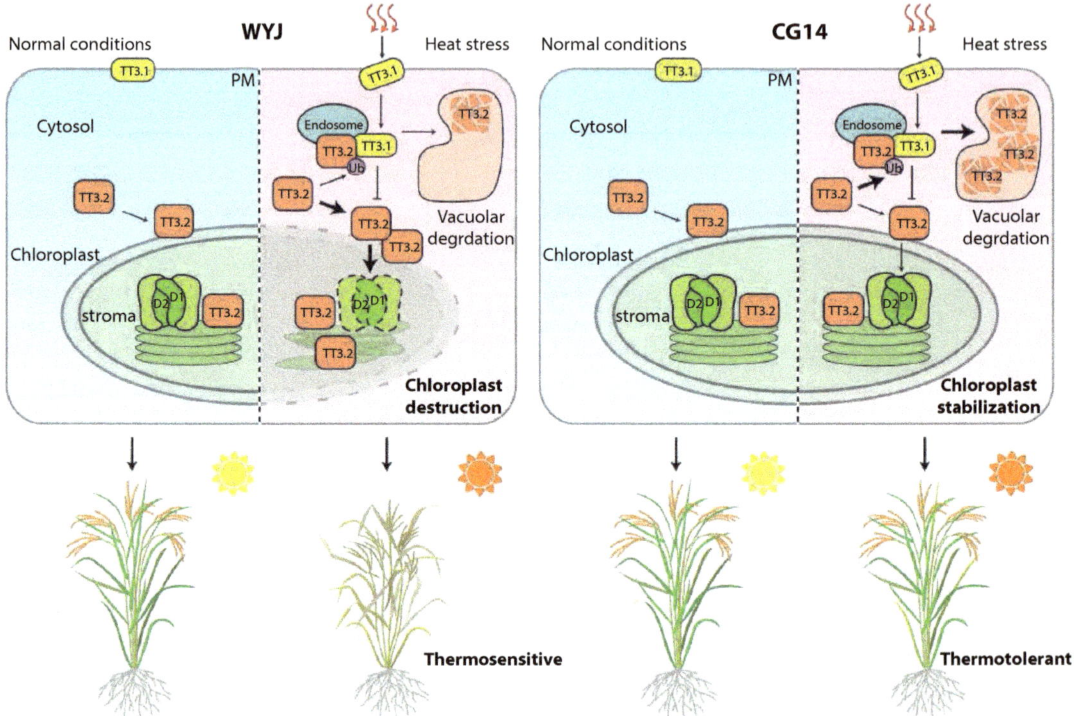

Figure 6.9 CRISPR/Cas9 Inserts Genes for Heat Tolerance

Certain proteins in the plasma membrane of plants, known as thermo-tolerance proteins (TT3.1), act as temperature sensors. When activated by heat, they enter the cell and bind to another protein (TT3.2) that normally interacts with chloroplasts. Within the cell, TT3.1 marks the second protein for degradation. A common variant, TT3.1(WYJ), has low activity, which allows the concentration of TT3.2 to build up and collect in the chloroplast, causing it to break down and the plant to wither and die. Another variant, TT3.1(CG14), has high activity, which keeps the concentration of TT3.2 low, preventing chloroplast breakdown and thus better heat tolerance. Using CRISPR/Cas9 to insert this gene variant in plants will enable plants to tolerate the higher temperatures of a globally warmer planet.

plant exhibiting early flowering, enhanced photosynthesis, and improved nitrogen use efficiency, resulting in a substantially increased yield in rice and wheat. Using CRISPR/Cas9 to insert additional copies of the genes for OsDREB1C in plants might yield higher productivity and a more robust food supply.

What about improving the vitality of plants in a warmer climate? Scientists have found two genes responsible for a plant's ability to detect and communicate temperature fluctuations to the cell. By comparing the heat-tolerant rice variant from Africa (CG14) to a typical Japanese rice variant (WYJ), labs in Shanghai and other locations in China were able to identify the two genes that led to heat tolerance in the African variant. As shown in Figure 6.9, one gene produces an enzyme, TT3.1, that typically resides on the cell's plasma membrane; and the second gene produces a protein, TT3.2, that is in both the cytosol and the chloroplast, where it binds to the intracellular membrane known as the thylakoid membrane. Researchers saw that when the heat stresses the plant, the two proteins come together inside the plant, and TT3.1 adds a degradation marker to TT3.2, marking it for destruction. In the Japanese variant, TT3.1 is less active, and TT3.2 builds us, causing the chloroplasts to fall apart and the plant dies. In the African variant, the TT3.1 still moves to the cytosol when the cell is heat-stressed, but it is more active, so more of the TT3.2 gets degraded, and the chloroplasts remain intact. These two genes are conserved in significant crops, such as maize and wheat, and are valuable resources for breeding highly heat stress–tolerant crops. Incorporating this variant by CRISPR/Cas9 could produce heat stress–tolerant plants and reduce grain yield loss caused by heat stress.

From Food Innovators: Reimagining Food Components and Production Methods

Continuing a Western diet is not environmentally sustainable or compatible with concerns for animal welfare and human health and well-being.[52] It will be possible to reach the goals of sustainable food production using plant-based products as alternatives to meat, fish, eggs, and dairy products.[53] In addition, support is needed for consumer education programs and food innovation grant programs.

Nutritious meat, fish, egg, and dairy analogs prepared from renewable food sources such as algae, insects, mycoproteins, and plants are currently available. Addressing challenges such as making these foods affordable, improving food processing techniques, and consumer reluctance are still necessary. Three examples of food innovations are plant-based and cultured meat products, printable edibles, and note-by-note cooking.

Plant-Based Meat Products

Long-term, large-scale production of the steaks and roasts that many of us love is incredibly energy-intensive, fraught with animal rights issues, and not sustainable in the long term. Plant-based meats are becoming increasingly successful in the search for satisfying alternatives that are more planet-friendly. Many consider eating plants a win-win-win: we diminish our carbon footprint and honor concerns over animal welfare while at the same time eliminating the health risks associated with red meat. Plant-based protein analogs such as textured vegetable protein (TVP) have been popular since the 1960s. They are commonly used to add nutritional value to chili and sloppy joes or as additives to Bolognese sauces. Their low cost and long shelf life have also made them a backbone of foods used in disaster relief, prisons, schools, and backpacking. Other plant-based protein sources are Seitan, made from wheat gluten, and Quorn, made from fungal mycoproteins. July 2016 saw the launch of the Impossible Burger by Impossible Foods.

The Impossible Burger is made from soy and potatoes flavored with allium (garlic, onions, shallots) and spiked with coconut and sunflower oils to make it grill like a burger, all bound together with materials such as methylcellulose and food starch. One final ingredient is an oxygen-binding protein, leghemoglobin, from the roots of the soy plant.[§] The name derives from "leg"—from a legume, as the soy plant is in the legume family—and "hemoglobin"—the heme-based protein similar to animal hemoglobin. The protein most closely mimics myoglobin, the oxygen-binding protein in animal muscle that colors the juices from steaks or burgers. In the Impossible Burger, the leghemoglobin adds to the meaty taste, the juiciness of the burger, and the blood red color. The novelty of this bleeding burger has begun to wear off, and consumers and fast-food restaurants are losing interest. A common complaint is that the plant fats do not provide the texture and flavor of the fats in the meats they replace. One solution to this problem is to culture the animal fat cells that make the fat and then introduce these cultured fats into plant-based meat substitutes like burgers and bacon.[54] This approach has led to some early success, and several companies are investing in this approach to bring consumers back to plant-based protein. The result is very satisfying, and products like this and others may represent the future of much of our protein consumption.

Cultured Meat Products

Another strategy to provide meat proteins while acknowledging concerns about the climate and removing ethical issues regarding animal welfare is cultured meat, which is meat grown outside of an animal, or ex vivo.[55] The idea that one might not need an entire animal to grow a steak or a chicken breast was first made popular in the 1950s by a Dutch researcher, Willem van Eelen.[56] Having known starvation and hunger as a prisoner in World War II, he was highly motivated to consider food security. By the 1970s, scientists had successfully cultured a guinea pig aorta to replace a human aorta in cardiac surgeries. The first U.S. patent for a cultured meat product for human consumption came along in 1991, and by early 2000, small quantities of foods were being produced in research laboratories. Cultured meat is produced from a particular kind of cell, called a stem cell or primary cell, that can be obtained from an animal without slaughtering it. As shown in Figure 6.10, the cells har-

---

§ Leghemoglobin is from a symbiotic bacterium in the root nodule of the plant, not the plant itself.

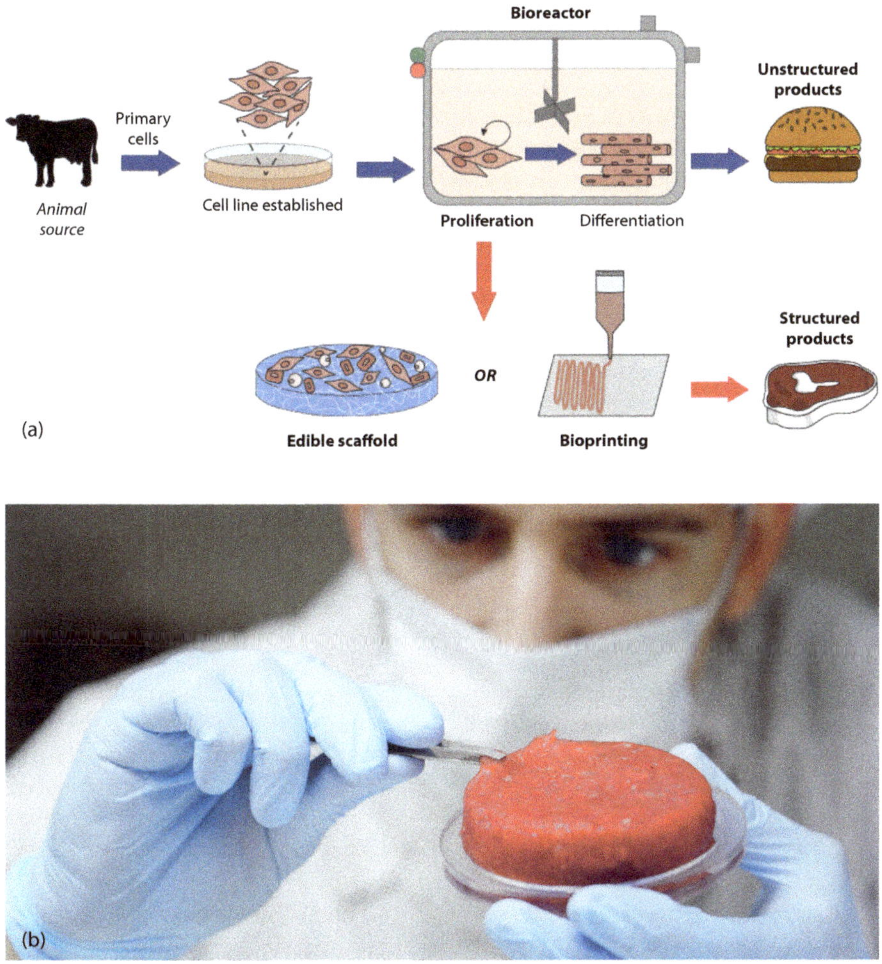

Figure 6.10 Culturing Beef Cells in the Lab

The flowchart in (a) shows the steps that lead from extracted primary cells from animals that can be established in a cell culture medium and then grown in a bioreactor to form both unstructured meat patties and structured meat products such as steaks. In (b), the unstructured cultured product is examined by a food scientist.

vested from the animal are then stabilized as a cell line in a petri dish. Later samples of cells from these cell lines are grown in a bioreactor in a broth containing hundreds of different nutrients and signaling molecules. The cells must first divide or proliferate before they differentiate—which causes them to mature into muscle fiber cells. Minced meats for burgers, patties, or meatballs can be harvested, cooked, and consumed. If the desired end product is a steak, it must be further organized to grow on an edible scaffold or bioprinted to incorporate the fat and muscle cells one expects from a well-marbled steak or ham.

Efforts to make such a product commercially viable have yet to lead to products available to the consumer that are competitively priced, but they are close. In 2013, Mark Post at Maastricht University announced that his lab cultured the first beef hamburger patty priced at just over $300,000.[57] In the decade since then, sixty start-up companies have been formed worldwide. As of this writing, twelve plants across the globe have started to produce cultured meat and fish products such as pork, beef, and salmon for consumers. A challenge to the industry at this point is the cost associated with the growth media for the cells—and this price will have to drop significantly before cultured foods become commercially competitive. Eat Just, a cultured meat company, produced its first cultured chicken nugget dinner for about $50. Today one can obtain a cultured chicken nugget dinner at its Singapore restaurant 1880 for about $17.

Printable Edibles: Digital Cooking

What follows might seem like an excerpt from a science fiction or future fantasy movie. Yet it is worth our while to consider one solution to the food waste concerns: a food printer borrowed from the 1960s futuristic cartoon, "The Jetsons."[58] These wild ideas might provide solutions that ensure a sustainable food future. Printable edibles or digital cooking is an innovative system that uses a 3D printer for preparing food.[59] Sustainable benefits of food printing are reduced carbon emission, reduced waste of raw materials, improved operations performance, more efficient energy consumption, no inventories, customization and easy manufacture, economic balance, and design movability. Individualized food components are derived from natural compounds, contained in shelf-stable source pods like ink cartridges, and mixed according to a recipe on a computer. The consumer selects a food item, punches buttons, and watches as the food is printed, cooked, and delivered ready to eat. Food printers were first introduced around 2007 to prepare foods such as confections that do not require additional cooking to be edible.

Today food printers are available for about $4,000 with a posted 2-week delivery time from companies such as 3D byFlow.[60] These printers are marketed to high-end restaurants and food producers. However, their vision for the future is to have such devices in every home. In 2021, the added adaptation of using lasers to cook food was shown to work. Excitement was generated by the early experiments reported in the Nature Partner Journal, *Science of Foods*.[61] Food is deposited in a specific 3D array, and the energy to cook the food is provided to mm precision by three lasers: infrared (wavelength 10.6 microns), near-infrared (980 nm), and blue (wavelength 445 nm). J. D. Blutinger at the

Figure 6.11 Food Fabricator of the Future

Concept rendering from the Columbia food lab of a food fabricator that can both create the food from ingredient packets following computer instructions and cook the food using lasers.

Creative Machine Lab at Columbia University believes that in the future most kitchens will have a new appliance, a 3DFP, which will be like a "personalized chef" that can interpret digital instructions to prepare and cook food on the spot. In a blind test, tasters were asked to compare the chicken prepared conventionally in an oven to homogenized chicken printed and cooked using a 3DFP. Most tasters preferred the laser-printed chicken, reporting that the 3D product was moister and had a more uniform texture. Hod Lipson, professor of mechanical engineering at Columbia, says the technology is here, and with a motivated business investment, these new 3DFPs could be in our kitchens in a year. A low-cost 3DFP with a capacity for ten different ingredient pouches might look like the concept-rendered appliance shown in Figure 6.11.

Note-by-Note Cooking

Note-by-Note Cooking (NbN) was first developed and popularized by Hevré This in the 1990s. In NbN cooking, foods are prepared by mixing pure compounds, such as potato starch, dehydrated milk powder, and flavor extracts, often in unexpected ways.[62] Molecular gastronomy also uses new techniques to prepare foods, but in contrast to NbN cooking, traditional food ingredients, such as potatoes, milk, and butter, are used. For NbN chefs, choosing the ingredients is relatively easy; putting them together is a challenge.

This new cooking movement addresses the global energy crisis, the concern for reducing food transportation costs, and the general improvement in human health in the following ways. NbN-centered food production deconstructs food into its components at the production site, fractionating the food into various pure components, removing the water, and producing the purified "notes," which are lighter and cost less to transport. When fractionation is done on site, the removed water can be returned for irrigation, and the wastes can be transformed into compost, plant, or insect food. It has been argued that given the on-site extraction of the pure compounds, food spoilage would be vastly reduced or eliminated. Nutritionally balanced foods can be prepared, thereby saving lives by reducing human diseases such as Type 2 diabetes, heart disease, and obesity. As a side benefit, food intolerances from gluten or allergies to eggs, peanuts, dairy, or seeds can be eliminated by removing the gluten or allergens from foods and, in so doing, removing a significant threat to some consumers. Since pure compounds are extracted, the origin does not matter. For example, a chlorophyll extract might come from algae and could be used safely in food for someone allergic to soy. More economically and agriculturally beneficial crops can be grown (in the case of chlorophyll, for example, algae might be easier to grow than spinach). Finally, food storage for these mostly dried products would not require electricity. So they could be shipped to more remote and impoverished sections of the world, thus reducing food inequalities.

Consumers may need to realize that many commonly used food ingredients, such as high fructose corn syrup, are already processed or deconstructed from existing foods. Water, salt or sodium chloride, sucrose, and gelatin are already fundamental notes prepared for us by the food industry. Many food thickeners, such as agar and calcium chloride, might be added. It is easy to imagine that additional sugars, polysaccharides, amino acids, and fats will be added to the introductory notes. The food additives industry already produces pigments, vitamins, preservatives, and flavors such as artificial vanilla.

Currently, recipes are being developed and courses utilizing the techniques of NbN cooking are being taught at Le Cordon Bleu, one of the premier cooking schools in the world. Figure 6.12 shows a few dishes prepared by NbN cooking, a breakfast dish that simulates egg, toast, and bacon but is made of deconstructed ingredients, and a unique creation named "Chick Corea" after the jazz pianist. Here is an NbN recipe for a sauce that might be used with this dish:

- Melt 100 g of glucose and 20 g of tartaric acid in 200 ml of water over low heat.
- Add 2 g of polyphenol
- Boil and add sodium chloride and piperine to taste

Figure 6.12 Note-by-Note Dishes from Synthetic Ingredients

Two dishes created by chefs at Le Cordon Bleu in Paris. A breakfast dish, top; bottom, a special dish named for the jazz fusion pianist Chick Correa.

- Bind the sauce with amylose
- Remove the preparation from the heat and stir in 50 g of triacylglycerol
- Serve warm over eggs benedict or vegetables

Until you realize that all these compounds are edibles—the tartaric acid and polyphenol are extracts from grapes, the sodium chloride is table salt, the piperidine is the pungent extract from black pepper, the amylose is a type of starch, and triacylglycerol is the major component of oil or fat—you might think that the directions for making the sauce are from a chemistry lab manual. Hevré This admits that "it will take a minimum of 10 years" for his ideas to catch on.[63] But that does not stop the explosion of new courses in culinary schools on NbN cooking and global competitions each year for new NbN dishes.

SUMMARY OF CHAPTER 6

In this chapter, we identified the relationship between food waste and climate change by examining the sources of greenhouse gas emissions from various sectors of the food production pipeline. We saw that scientists have identified food choices we make daily, such as a heavily meat-based diet, that can be changed to reduce these threats. Educating consumers about planet-friendly food choices is already a priority at the FDA, in schools, and in the media. Meanwhile, scientists and CEOs have sought ways to reduce food spoilage, label foods better, reduce packaging, reframe what we mean by "quality," and find better ways to distribute foods to address the problems of food scarcity and food insecurity. Looking ahead, we see even more hope from scientists from many disciplines.

Material scientists are reimagining ways to reclaim, recycle, and upcycle our waste foods. Chemists and biologists are using CRISPR/Cas9 protocols to design new drought-resistant plants that mature more rapidly and require fewer fertilizers and pesticides. Other food innovators are developing plant-based meats, tissue-cultured meats, and printable foods and using note-by-note cooking to reimagine and provide food using new materials and methods.

PART

IV

PERFORMING CULINARY EXPLORATIONS

In the kitchen and laboratory, understanding is developed by experimentation and practice. In the following exercises, you will be guided through a series of experiments that will allow you to see how food can be transformed using the techniques explored in the preceding chapters. One of the most important things about being in a science laboratory is to be trained in safe practices and the safe storage, handling, and disposal of materials. As a prelude to the detailed culinary experiments, we will present guidelines for keeping you and your consumers safe. Each worker must be certified through Safe Serve[1] training courses and exams to prevent food-borne illness in the food service sector.

Chapter 3 explored the many cooking techniques that involve the transfer of heat either into or out of the ingredients for a dish. We present two experiments that show how this works. **Experiment 1: Transforming Food with Heat** lets you explore the many ways the composition of an ingredient, here a simple egg, determines how it will respond when heat is added and compares heat transformation to several other methods. In a cool contrast to traditional cooking by adding heat, the next experiment, **Experiment 2: Transforming Food by Removing Heat—Cryogenic Cuisine**, explores how to create delicious dishes by this method.

Chapter 4 explored how extremes of pH, pressure, and microbial action can transform foods. In **Experiment 3: Transforming Milk to Cheese and Yogurt**, you will get a chance to see how both pH and microbial activity can be used to create incredible new tastes and textures, starting with basic milk. In **Experiment 4: Sous Vide—Cooking at Low Pressure without Oxygen**, you can experiment with this efficient and straightforward method of slow cooking that allows superior flavor and texture development by excluding oxygen and preventing problems due to uneven heat distribution or overcooking.

Chapter 5 explored how you can create delicious foods by transforming them into colloids or emulsions by adding mechanical energy. In **Experiment 5: Vegetable Foams**, you will create two sweet or savory foams, one made from potatoes and the other from squash. In **Experiment 6: Spherification**, you can experiment with stabilizers to create small spheres used to create mocktails and reverse spherification to make small fruit-filled globes of flavor that will be presented as amuse-bouches.

The quality of scientific work is evaluated by peer review; the results and conclusions are sent to a panel of experts who decide if the work meets specific scientific standards. In a kitchen, the success or failure of a dish is evaluated by those who taste it. Several experiments contain "Culinary Challenges" in which judges are asked to assess your dishes. They identify the dish that best incorporates many techniques and appeals to their sense of taste, aroma, plating, and creativity. We recruit judges from interested colleagues who undergo training specific to each lab. These challenges will motivate your team to think more broadly and creatively and to be accountable for your culinary decisions. The challenges may become the favorite part of your lab.

CHAPTER 7

The Culinary Laboratory

SIX EXPERIMENTS ARE INCLUDED IN LATER PAGES IN THIS CHAPTER:

- Experiment 1: Transforming Food with Heat
- Experiment 2: Transforming Food by Removing Heat—Cryogenic Cuisine
- Experiment 3: Transforming Milk to Cheese and Yogurt
- Experiment 4: Sous Vide—Cooking at Low Pressure without Oxygen
- Experiment 5: Vegetable Foams
- Experiment 6: Spherification

Before getting into the specific directions for each of the experiments, it is important to review safety considerations in the laboratory. These guidelines will help keep you and your consumer safe. One way these experimental procedures are more than just recipes is that each experiment has been designed with a learning goal in mind that is relevant to food chemistry. The best way to demonstrate mastery of the learning goal is to collaborate with others to write a lab report. The basis and evaluation metrics for these reports are outlined here. While many formats will work fine, we have found the best tool for presenting and sharing these reports is through collaborative online Lab Wikis that are filed on Moodle, our class's learning management system.

INTRODUCTION AND EXPECTATIONS

Safe behavior in the laboratory is important, and in a laboratory where the products of your reactions will be consumed, it is even more important. Some of you might be familiar with labs, but it might be your first time in a lab-based environment for others. Here are the five most important things you must do to prepare for the best culinary laboratory experience.

1. COME PREPARED You must read the lab handout each week beforehand. To ensure that you are prepared, a pre-lab quiz will be given. Your time in the lab will fly by, and you and your lab partners need to be prepared and organized. In consultation, your lab group should assign roles for work in the lab and with the Lab Wiki. Three to four people are usually assigned to each group. While the exact role designations should change week to week, typical assignments are head chef to coordinate activities, sous chef to help with the preparation of ingredients, procurement chef, and, finally, a recorder to keep records and document the experiment with photos or videos that can be used in the Lab Write-up. The role of the procurement chef is to gather equipment and supplies, keep the work area clean, and have everything ready BEFORE the chefs begin to cook. This task is so essential that in the culinary kitchen, there is a particular term for it, "mise en place."

2. DRESS APPROPRIATELY No sandals, open-toed shoes, shorts, or sleeveless tops. Jeans, T-shirts, and sneakers are a perfect uniform for the lab. Long hair should be tied back. ServSafe guidelines require head coverings. We wear baseball caps.

3. WASH YOUR HANDS THOROUGHLY AND OFTEN Washing your hands before preparing food is essential. Use soap and water, and keep your hands in the flow of water for at least a minute, rubbing the soap through your hands and wrists and rinsing thoroughly. Our first laboratory meeting includes a demonstration of a careful handwashing technique. We first rub a fluorescent powder on our hands, wash them, and then use a UV light to check if the powder is gone. Handwashing must be repeated until the powder is gone. We may ask you to wear gloves for certain parts of the lab. When we do this, it does not mean that you do not have to wash your hands.

4. PAY ATTENTION You will work with sharp knives, hot pans, frying oils, and simmering liquids. Keep your cooking areas orderly and uncluttered. Note the locations of fire extinguishers at each workstation and notify the lab instructor or one of the teaching assistants if anyone gets burned or cut. Note the location of the first aid kit, which contains Band-Aids and gauze for minor cuts. In the event of a fire, everyone should evacuate the room and activate a fire alarm if available, and emergency response can be initiated by dialing 911 (or your local equivalent). If someone is injured, call 911 and stay with them until help arrives. Always notify your instructor.

5. CLEAN UP AFTER YOURSELF Your group finishes when your work area is as clean as when you arrived. All pots, pans, dishes, knives, cutting boards, blenders, etc., should be back in the workstations—cleaned, dried, and ready for the next group to use. Should an item be lost or broken, report it to one of the teaching assistants or your instructor.

Food Allergies and Preferences

All students must inform their instructors of food allergies, dietary restrictions, and preferences. When preparing food for the judges during culinary challenges, be sure to inquire if the judges have any allergies or dietary restrictions and plan your work accordingly. The eight major allergens identified by the FDA are milk, eggs, fish, shellfish, tree nuts, peanuts, wheat, and soybeans.

Handling Food Waste

As discussed in Chapter 6, food waste is one of the leading causes of greenhouse gas emissions that lead to climate change. Every effort should be made to reduce food waste in the Culinary Labs. Compost bins and trash containers for recyclable and landfill waste should be available. If it is not possible to use reusable tableware, disposable tableware such as forks, spoons, knives, plates, cups, and bowls can be made of biodegradable materials (polylactic acid). Pantry items (rice, flour, oils, spices) can be labeled with a purchase date and stored tightly sealed to ensure longevity. Every effort should be made to ensure that leftovers are not thrown away. Reusable storage containers can be used to allow students to bring ingredients and extra cooked foods home or to their residence halls.

Evaluation and Wiki Template

Student work and understanding are evaluated in individual pre-lab quizzes and a collaborative online lab report presented as a Lab Wikis. The Wiki emphasizes the broad reach of cooking into the liberal arts and includes sections on science, history, art, and literature and a creative element. Students work in groups of three or four, and each member takes responsibility for one of the unique sections below. All group members must contribute to the summary reflection. The individual roles rotate with each write-up. Winners of culinary challenges receive a small prize. Explicit descriptions of the expectations for the content of each unit are shown in Table 7–1 and summarized here.

- Unit 1: SCIENCE unit summarizes scientific background and—this is not just a recapitulation of the lab handout. It is required for each WIKI.
- Unit 2: RELATIONSHIP TO SYLLABUS—this unit ties the experiment to the readings and should be referenced to it. It is required for each WIKI.
- Unit 3: CREATIVE/MEDIA—this optional unit uses your creativity with writing, media, or the arts.
- Unit 4: SOCIAL SCIENCES/HUMANITIES—this optional unit pushes deeper into the sociological, anthropological, historical, economic, or nutritional aspects relevant to each lab.

Table 7–1 Components and Instructions for the Laboratory WIKIs

Unit 1—Science (required)	• For example, in the Egg Lab, Transforming Food with Heat, you can present information about the biology or chemistry of egg proteins and how each denaturant caused the proteins to denature. You should have collated the results on the denaturation using various acids and salt from your classmates. • This section might include molecular structures, flow charts, bar graphs, or other numerical or quantitative outcomes from your work. Be sure to provide citations and authorship.
Unit 2—Relationship to Syllabus (required)	• Each post must pick up on a theme that integrates the lab with the readings, films, or class visitors. For the Egg Lab, here are some examples of how you might tie the lab back to the syllabus: What is the origin of the contemporary fascination with sous vide eggs? Are eggs considered a major allergen? Is food safety compromised when eggs are left at room temperature? • 300 words max. Your work must include citations and authorship.
Unit 3—Creative/ Media (optional)	• Each post might contain a creative element (i.e., photo, illustration, video interview, poem, haiku, critique of a related piece of fiction, short story, recipe) relevant to the lab. You might wish to cook a dish and document its preparation and presentation with photos and a narrative. You could also provide a personal story that relates to this lab. • Photographs, drawings, or any other type of illustration must be accompanied by a narrative of less than 300 words to describe the work and its relevance to the topic. Include authorship.
Unit 4—Social Sciences/Humanities (optional)	• Each post **could** contain a section that picks up on a historical, ethical, anthropological, nutritional, or other aspect of the lab. • For example, in the Milk Lab, you could comment on which cultures consume fermented milk and why that might be. I found the story of how religious persecution of Sephardic Jews first in Spain, then Greece, and finally France was in part responsible for the appearance of yogurt in the U.S. • Be sure to provide citations and authorship. 300 words max.
Unit 5—Reflective Summary (required)	• Here is where you get to read through your Lab Group's WIKI in its entirety, keep tabs on what your classmates are doing, and reflect on what the group as a whole has produced. • Each group member should comment individually.

- Unit 5: SUMMARY REFLECTION—this required unit is a space for each group member to provide a final brief commentary after reviewing the work of their lab partners.

EXPERIMENT 1: TRANSFORMING FOOD WITH HEAT

GOAL: In this experiment, you will try different methods of transforming an egg by applying heat and observing how adding acid, salt, and base affects the outcome of cooking.

Background

Humans have eaten eggs for thousands if not hundreds of thousands of years. Eggs provide a protein-rich food, naturally preserved and readily abundant.

The anatomy of an avian egg will help us understand what is happening as we cook it.[1] A relatively rigid but semiporous shell protects the fragile interior. An eggshell is made from calcium carbonate, the same chemical that makes up chalk and limestone. The color of a shell depends mainly on the species of chicken. The interior has two primary compartments, the white and the yolk. The white of the egg is typically at a pH of 8.8. It comprises 90% water and 10% egg proteins such as albumin and avidin, which act as natural preservatives, and globulin and transferrin. By contrast, the egg yolk typically has a pH of 6.5, with 50% water, 20% protein, and 30% fat, which makes it very rich in taste. The yolk contains fats, cholesterol, vitamins, and minerals such as iron, vitamin A, vitamin D, phosphorus, calcium, thiamine, and riboflavin. Proteins in the yolk include low-density and high-density lipoprotein (LDL and HDL) and phosvitin, a highly phosphorylated protein essential in bone growth and in binding to calcium. The color of the yolk is primarily due to the molecule lutein, which is yellow, but the yolk's color can be influenced by diet.

Proteins are essential for understanding the structure of cooked eggs. When the egg is raw, the proteins are folded up, and the egg white is transparent, though it may be a bit cloudy. As discussed in Chapter 3 and depicted in Figure 3.12, when heat is applied to the egg by cooking, the egg proteins unfold, stick together, and form a network, resulting in a semisolid opaque egg white. The unfolded and aggregated proteins form the matrix that holds the water (as steam when cooking) and helps make cooked eggs fluffy. With continued cooking, more water is driven off, and the egg becomes dry and rubbery. Salt and fat help hold the matrix together and make eggs fluffy. The flavor of an egg depends on its freshness, the species, and the feed provided to the chicken.

Part A: Each lab group must do Part A-1: Effect of Acid and Salt on a Scrambled Egg and record your observations and share them with the class. The class should compile their results for this section as this will form the basis of your Lab WIKI's Unit 1: SCIENCE. Each group should also observe, take notes, and, when appropriate, taste. Parts A-2: Eggs Pickled in Vinegar; A-3: Emperor Egg; and A-4: Century Egg require only observations and notes.

Part B: Culinary Challenge. Each group will choose a basket from the front of the room. The baskets will contain equipment and ingredients for your group to do one of the following: B-1: Sunny Side Up Eggs, B-2: Soft-Boiled Eggs, B-3: Perfect Scrambled Eggs, or B-4: Poached Eggs. Each basket will contain herbs, spices, and other ingredients suited to your dish. A table of shared supplies (butter, milk, oils, condiments) will also be in the front of the room. You will have 30 minutes to prepare your dishes. At the end of 30 minutes, one member of your group will present the dish to a panel of judges, who are trained volunteer members of our community. They will taste and evaluate your dishes for flavor, appearance, creativity, and presentation, and after consultation, they will select the most successful dish. This group will receive a small prize.

Name	Structure
Lemon juice—citric acid	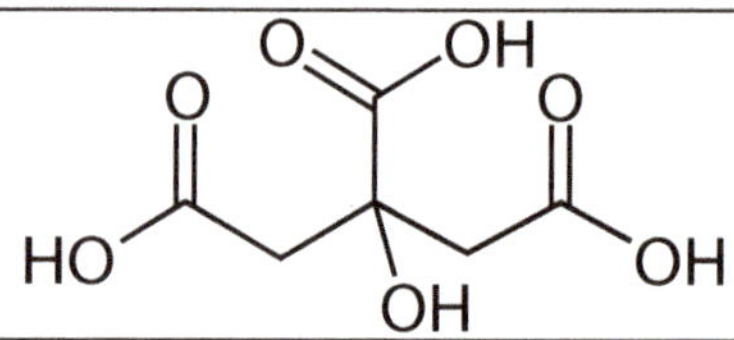
Vinegar—acetic acid	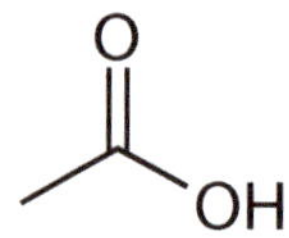
Cream of tartar—potassium bitartrate	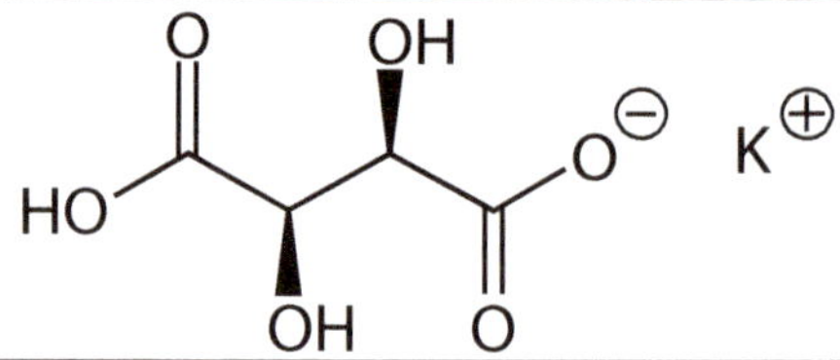
Salt	$Na^+ Cl^-$

Figure 7.1 Molecular Structures of the Ingredients Used to Denature Eggs for Experiment 1.

Part A. Eggs, Salt, and Acid

Various denaturants will be added to a raw egg before cooking, and the products will be compared to each other and to a control egg. The molecular structures of the three different acids and salt are shown in Figure 7.1.

Part A-1: Effect of Acid and Salt on a Scrambled Egg

GOAL: In this experiment, you will observe how treating a raw egg with different denaturants (citric, acetic, and tartaric acid and salt) affects how it cooks and the flavor.

INGREDIENTS

Each group should take two eggs and one of the four additives. Since pans and heat sources differ, each group will use one egg to do a control experiment in which the egg is just scrambled, with no additives.

INSTRUCTIONS

1. Add 1 teaspoon vegetable oil to a small skillet and heat on medium for about 2 minutes.

Table 7–2 **Data Sheets for Experiment 1 Part A**
Class Observations of Results of Various Egg Denaturants

	CONTROL	+ VINEGAR	+ CITRUS ACID	+TARTARIC ACID	+SALT
Color					
Homogeneity—distribution and size of bubbles (if present)					
Texture—fluffy or crumbly or rubbery					
Taste					
Other					

2. Meanwhile, crack an egg into a small bowl and mix with a fork until the yolk is thoroughly combined with the white. Do this once for the control egg and once for the treated egg.
3. For the control egg, skip this step. For the treated egg, add 1 tablespoon of either lemon juice, vinegar, 1% solution of or cream of tartar or 1/4 teaspoon of salt. Mix well. For the salted egg, wait 10 minutes. Vegan plant-based egg substitutes such as JUST-Egg can be used.
4. Add the egg mixture by pouring it into the middle of a small skillet, then swirling the mixture to cover as much of the skillet's surface as possible.
5. Cook the egg for 3 minutes, and note the fully cooked egg's texture, color, and appearance. Flip the egg to observe the underside as well.
6. Compare your results with those of the rest of the class*.
7. Be sure to wipe out the pan in preparation for the experiment in Part B.

Part A-2: Sample Eggs Pickled in Vinegar

GOAL: In this experiment, you will record your observations of the appearance and flavor of an egg cooked in a shell, peeled, and then marinated in an acidic solution known as a brine.

* Write your results on the data sheet in Table 7–2 that can be shared with the class. Be sure to collect the complete data set before you leave.

These pickled eggs have already been prepared for you. Be sure to inspect one and complete your observations on the data sheet. Tasting is optional.

INGREDIENTS FOR THE BRINE

- 2 medium onions, chopped
- 1 tablespoon pickling spices
- 1 teaspoon salt
- ⅓ cup white sugar
- 4 cups vinegar
- 12 hard-cooked eggs (boiled 10–12 minutes), shelled and left whole

INSTRUCTIONS

1. Place onions, spices, salt, vinegar, and sugar into a saucepan and boil for 5 minutes.
2. Pour the boiled liquid over a dozen peeled hard-boiled eggs and place in a sterilized jar and seal.
3. Leave on the counter overnight.
4. Refrigerate; the longer they are allowed to sit, the better they will taste.

Part A-3: Inspect the Emperor's Egg

GOAL: In this experiment, you will record your observations of the color, texture, aroma, and appearance of a raw egg soaked in an acidic solution for an extended period. As the egg is raw, do not taste it.

This egg has already been prepared for you. A raw egg has been left in a vinegar solution (5% acid) for at least two weeks. What do you see? What is missing, and how can you explain what has happened? Fill out the data sheet with your observations. Why do you think it is called an emperor's egg?

Part A-4: Inspect and Taste the Century Egg

GOAL: In this experiment, you will record observations of the appearance and flavor of a century egg treated for an extended period with base.

Whether you call it century egg, hundred-year egg, or millennium egg, this Chinese delicacy dates back centuries to the Ming dynasty. While you might think it takes centuries or millennia to make these eggs, the procedure takes about a month.

Traditionally, century eggs were made by covering raw chicken or duck eggs with salt, lime, and wood ash, wrapping them in rice husks, and waiting several weeks.[2] During this time, the pH of the egg rises, essentially cooking the egg with base. The chemical process breaks down some proteins and fats into smaller, more complex forms. Modern methods will use lye (sodium hydroxide) to treat the egg.

Century eggs might be locally available from an Asian food market or online. The unshelled egg is shown in Figure 4.4c in Chapter 4. A delicious way to serve a century egg is a traditional Korean style accented with creamy tofu, sesame oil, and sliced scallions. Take a portion back to your cooking station to inspect it and record your observations for both the intact and peeled egg. Comment on the color, consistency, odor, flavor, or any other aspect of the century egg that you find notable. Those curious to make their own can follow readily available online instructions.[3]

Part B: The Culinary Challenge

GOAL: The culinary challenge helps you develop your creativity and demonstrate mastery of the techniques. It will broaden your culinary repertoire, introduce you to new and perhaps unfamiliar ingredients, and invite you to create pairings of ingredients from your food basket.

In this part of the lab, we present four methods of cooking eggs. Each group will make a blind choice of one of the baskets, and at a signal, all groups will open their baskets to discover which egg recipe they will be making. Your group's task will be to cook the eggs using the assigned method and add additional elements from your basket at your discretion. You will have a total of 30 minutes. You will present your dish to the judges at the end of that time. Since the egg dishes for the challenge are assigned randomly,* you should read *all* the instructions before coming to the lab. The tasting judges will choose the winning dish based on its taste, aroma, appearance, creativity, and use of technique. Winners receive a small prize.

Part B-1: Sunny Side Up Eggs

(Perfect white with edges a bit crispy; yolk creamy and a bit runny)

INGREDIENTS

- 4 cracked eggs (two each into two small bowls).
- 2 teaspoons vegetable oil

* Vegans in the lab will not be assigned randomly but can participate by being assigned Part B-3: Scrambled Eggs and using the plant-based substitute, JUST-egg.

- 2 teaspoons butter (cold) cut into four pieces
- salt and pepper
- basket ingredients

INSTRUCTIONS

1. Use 4 cold eggs (so the yolks are more difficult to break). Cracking two eggs into each of two little bowls helps in the delivery. Add salt and pepper.
2. Add 2 teaspoons vegetable oil to a 12-inch skillet and heat to medium-high. After 3 minutes, add butter and swirl to melt, mix, and coat the skillet.
3. Carefully add eggs and immediately cover the skillet. Let it sit on the heat for one minute, and remove it from the heat. Use the timer.
4. Finish cooking off the heat with the lid on (15 seconds to 2 minutes, depending on the desired runniness of the yolk).
5. When checking the eggs for doneness, lift the lid just a crack to prevent loss of steam should they need further cooking.
6. When perfectly cooked, the thin layer of white surrounding the yolk will turn opaque, but the yolk should remain runny.

To cook only two eggs, use an 8- or 9-inch nonstick skillet and halve the oil and butter. You can use this method with extra-large or jumbo eggs without altering the timing.

These serve up nicely with some toast grilled alongside to perfection.

Part B-2: Soft Boiled Eggs (Fool-Proof Method)

INGREDIENTS AND EQUIPMENT

- 4 eggs cold from the fridge
- 4 cups and spoons
- topper
- basket ingredients

INSTRUCTIONS

1. Set the burner to a high setting when you begin this experiment. Fill the pot with water to a depth of only ½ inch. Bring to a boil.
2. Set the timer for 6 ½ minutes. Add large eggs that are unblemished and cold. Cover the pot. Cook for exactly 6½ minutes.
3. Remove from heat, rinse in cold water, and serve with salt and pepper.

For easy eating, soft baked eggs are usually presented in an eggcup with the top "cap" of the eggshell removed. Cups and toppers are commercially available,[4] but they are not necessary for preparing delicious soft-boiled eggs. Use the egg topper provided to score and remove the top of the eggshell. The topper works by grabbing the top cap of the eggshell. To operate, put the egg in the cup and place the topper on top of the egg. Pull the black plunger up and release. This single stroke should produce a crack outside the shell, facilitating a clean peel to reveal the soft egg inside.

The primary heat source used to cook the egg comes from the steam, which stays at a constant temperature of 100°C (212°F). This method is more reliable than submerging the eggs in water because the chill of the cold eggs can lower the water's temperature, making the cooking time less sure and depending on the number of eggs being cooked.

Part B-3: Perfect Scrambled Eggs

(Scrambled eggs cool quickly, so cooking them just before plating is essential.)

INGREDIENTS

- 4 large eggs plus one additional yolk
- ¼ teaspoon table salt
- ground black pepper
- ¼ cup half and half
- ¾ tablespoon unsalted butter
- basket ingredients

INSTRUCTIONS

1. Crack eggs into a medium bowl. Add the additional egg yolk. Add salt, pepper, and milk. Whip with a fork until the streaks are gone and the color is pure yellow; stop beating while the bubbles are still large. Do not overbeat, or your eggs will be rubbery.
2. Set the timer for 2 minutes before you start, but do not yet push the start button. Prepare your serving dishes and utensils ahead of time.
3. Put ¾ tablespoons butter in a 10-inch nonstick skillet and set the pan over medium-high heat. When the butter foams, swirl it around and up the sides of the pan. Do not let the butter brown. Before the foam completely subsides, pour in beaten eggs. With a wooden spatula or a nonstick-safe egg turner, push eggs from one side of the pan to the other, slowly but deliberately, lifting and folding eggs as they form into curds until the spatula leaves a trail in the beaten eggs. Then lower the

heat and fold the cooking eggs for another 30–60 seconds until the eggs are nicely clumped into a single mound but remain shiny and wet, 1½ to 2 minutes. Serve immediately.

For eight eggs plus two yolks, season with ½ teaspoon salt, 4 grinds of pepper, and ½ cup half-and-half. Heat 1 tablespoon of butter and use a 10-inch skillet. Cooking time is about 1 minute.

Scrambled eggs are the quintessential American breakfast. You may find some bread and other adornment in your basket to plate this dish to perfection.

Part B-4: Poached Eggs

Poaching is a culinary technique in which food is cooked by submerging it in a hot liquid below the boiling point. Poached eggs are at the heart of eggs Benedict, where the egg sits atop Canadian bacon or ham on an English muffin with hollandaise sauce.

INGREDIENTS/EQUIPMENT

- 4 eggs
- large pot with a cover
- long slotted spoon
- vinegar
- salt and pepper
- basket ingredients

INSTRUCTIONS

1. In a large pot, fill water until the level reaches 2 inches. Cover the pot and place it on high on the burner.
2. Once the water boils, turn off the heat and add 1 teaspoon of vinegar.
3. Crack 1 egg into a small bowl.
4. Adjust the burner to bring the heating level back to medium and keep the water hot but below boiling. There should be minimal vigorous bubbling.
5. Use a spoon to create a vortex in the pot by swirling the water slowly for a few rounds.
6. Before the water movement stops, gently slide the egg from the bowl into the hot water.
7. A second egg can be gently added by this method. Cover the pot, and let it simmer on medium heat for 3 minutes. Simmer for 2 more minutes if you want a firmer yolk.
8. Scoop the egg(s) with the spoon and sprinkle salt and pepper to taste.

EXPERIMENT 2: TRANSFORMING FOOD BY REMOVING HEAT—CRYOGENIC CUISINE

GOAL: In this experiment, you will gain experience working safely with several cryogens and observe and record how food changes when heat is removed.

Background

We have used heat to "cook" food, such as eggs, under various conditions. In this lab, we will explore how cold temperatures can be used to prepare foods in ways that might surprise, shock, and satisfy you and your taste buds. Nathan Mhyrvold, author of *Modernist Cuisine*, writes:

> Since man's discovery of fire, cooking has been mainly a process of subjecting food to high temperatures that chemically alter its color, taste, and texture. But the invention of cryogenic technology has handed chefs an exciting new tool—liquid nitrogen—for transforming food in fun and surprising ways. In our culinary laboratory, we use this ultra-cold liquid to cryo-poach oils, cryo-shatter cheese, cryo-powder herbs, and cryo-grate meat. It is great for making instant ice cream and perfectly cooked hamburgers. For many years, the coldest substance chefs had ready access to was dry ice (frozen carbon dioxide, CO_2), which sublimates directly to CO_2 gas at -80°C (-109°F). Although dry ice has some interesting culinary uses, its solid form limits its utility. Liquid nitrogen boils at a far colder temperature: -196°C (-321°F), about as many degrees below zero as hot fryer oil is above zero. And because nitrogen is stable as a liquid (unlike dry ice), it is relatively easy to store the liquid and pour it over food or into a bowl. Because its viscosity is about one-fifth that of water and it has relatively low surface tension, liquid nitrogen flows rapidly into nooks and crannies in foods, such as hamburger patties, that have rough or irregular surfaces. The cooks at our lab use it to make fantastic burgers that are first slow-cooked to medium rare, then dunked briefly in liquid nitrogen to freeze a thin layer of the exterior and, finally, deep-fried. The deep-frying creates a perfect brown crust and thaws the frozen layer but does not overcook the interior. Speed is crucial for freezing foods without damaging their texture. In general, the faster the freezing process, the smaller the ice crystals and the less they disrupt the cellular structure of the food. Since the 1970s, chefs have used liquid nitrogen to make super smooth ice cream. More recently, chefs have started using it to flash-freeze delicate foods such as foie gras. Because liquid nitrogen is a relatively new addition to the kitchen, many other applications of this versatile fluid still await discovery.[5]

Part A: "Sugar on Snow"—Comparison of Three Cryogens

GOAL: You will compare how maple syrup responds to different chilling methods to make "sugar on snow," a maple candy.

INGREDIENTS/EQUIPMENT

- 3 covered ice buckets or Styrofoam coolers with lids
- 3 Styrofoam bowls

- insulated gloves
- goggles
- large mixing spoons
- digital thermometer for heated syrup
- small pot of shaved ice or snow
- dry ice
- liquid nitrogen
- 1 cup pure maple syrup
- vegetable oil to rub on the edges of the pot
- pickles as palate cleansers

CAUTION: Liquid nitrogen and dry ice can cause burns if not carefully handled with insulated gloves and eye protection. Under NO CIRCUMSTANCES should you attempt to consume liquid nitrogen. Attempts to do so have resulted in severe injury.[6]

INSTRUCTIONS[7]

1. Obtain samples of each of the three cryogens: shaved ice or snow, finely ground dry ice, and liquid nitrogen (dispensed by an instructor or teaching assistant as needed). Use the storage containers provided. Review the safety video regarding the safe handling of liquid nitrogen: https://www.youtube.com/watch?v=Id_OC8qsH6E.
2. Pour 1 cup of pure maple syrup into a medium saucepan. Grade A medium amber is the best grade for candy making.
3. Use a paper towel to coat the pot's rim with a small amount of vegetable oil so the syrup will not boil over, and keep the boiling liquid low in the pot.
4. Carefully heat the syrup in the pot over medium heat, monitoring the temperature of the syrup while it is boiling. Foaming can be controlled by stirring the syrup with an oil-coated spoon. The syrup is ready when it reaches 119°C (246°F).
5. Remove "tempered" (heated and thickened) maple syrup from the heat source and gently pour it into a smaller Pyrex measuring pitcher. Cool briefly (5 minutes).
6. Carefully fill the Styrofoam bowls half full of cryogen while wearing the personal protective equipment (PPE), insulated gloves and goggles.
7. Without stirring the syrup or agitating the pitcher, gently pour about ⅓ of the tempered maple syrup over shaved ice or snow in thin ribbons (i.e., not in a big blob).
8. Gently pour ⅓ of the tempered maple syrup over the dry ice.
9. Pour the remainder of the tempered maple syrup into the liquid nitrogen.

Table 7–3 **Data and Tasting Sheets for Experiment 2 A—Cryogenic Cuisine**

Cryogen	Molecular Structure of Cryogen	Temperature of Cryogen	Phase Change	Comments
Ice				
Dry ice				
Liquid nitrogen				

Cryogen	Appearance and Clarity	Elasticity . . . Brittleness	Taste	Comments
Maple sugar on snow				
Maple sugar on dry ice				
Maple sugar on liquid nitrogen				

10. Compare the three candy products and fill in the data sheet in Table 7–3. Let the liquid nitrogen candy warm up before you taste it, or you may frost your tongue. Pickles can be used in between tastings as a palate cleanser.

Part B: The Culinary Challenge—Creating a Cryo-Appetizer

GOAL: The culinary challenge helps you develop your creativity and demonstrate mastery of the techniques. It will broaden your culinary repertoire, introduce you to new and perhaps unfamiliar ingredients, and invite you to create pairings of ingredients from your food basket.

Each group will blindly choose a basket containing 4 to 5 surprise ingredients to create your cryogenic appetizer. You will collaborate with your teammates to prepare an appetizer in 30 minutes using the ingredients provided and as many of the cryo-techniques as possible (**cryo-poaching** oils, **cryo-shattering** cheese, **cryo-powdering** herbs, and **cryo-grating** meats). Plate one serving for each judge for evaluation and authorize one team member to present the dish. Your food will be judged by its taste, texture, and appearance on the plate. Remember to consider how you will keep your appetizer cold while it awaits tasting. It isn't necessary to plate *all* the food. Keep some for your own consumption.

Optional Demonstration of Liquid Nitrogen Ice Cream and Sorbet

Dean's Ice Cream

A recipe that the author has field-tested on more than a thousand college students.

INGREDIENTS

- 2 quarts half and half
- 1 quart light or heavy cream
- 1 cup sugar (to taste)
- 1 tablespoon vanilla
- 5 liters liquid nitrogen

EQUIPMENT

- personal protection equipment (gloves and goggles)
- 5-gallon stainless steel mixing bowl
- sturdy large spoons and ladle
- dewar for liquid nitrogen
- bowls, spoons, cones, and sprinkles for distribution

INSTRUCTIONS

Note: Fresh fruit sorbet can be substituted for the ice cream mixture.

Mix the liquid ingredients with the sugar to blend in a large 5-gallon stainless steel bowl. More sugar can be added to taste. Once the sugar is dissolved, the liquid should be stirred continuously as the liquid nitrogen is added SLOWLY until thickened (about 5 minutes). Serve immediately.

EXPERIMENT 3: TRANSFORMING MILK TO CHEESE AND YOGURT

GOAL: You will observe some of milk's colloidal properties with the addition of lemon juice to make ricotta cheese and the addition of a starter culture to make yogurt.

Table 7–4. Nutritional Composition of Cow's Milk

Nutritional Component	Average % by Mass
Water	87.0
Solids	13.0
Lactose sugar	4.8
Fat (whole milk)	4.0
Protein	3.4
Minerals	0.8

Background—The Chemical Makeup of Milk

Milk is a colloid whose nutritional composition, as shown in Table 7–4, can vary considerably depending on the species (cow, goat, sheep, etc.) and the animal's health. The last four nutrients are dissolved or suspended in the water.

Cheese is a milk-based food produced by coagulating the milk protein **casein**, from which the name "cheese" is derived. The main ingredients in cheese are proteins, fat, and water. Milk can be obtained from many domesticated mammals, such as cows, buffalo, goats, or sheep. During production, the milk is usually acidified, which causes coagulation or unfolding and then aggregation of proteins. The solids (curds) are separated from the liquid (whey) and pressed into final form. Unique flavors can be derived from each breed, the types of foods that the animal eats, or even during processing from the molds on the rind, the outer layer, or throughout.

Cheeses have a greater resistance to spoilage than milk, which is one of the likely reasons that early civilizations developed the art of cheese-making. The hardness or softness of a cheese will often dictate the shelf-life, with harder cheeses such as parmesan lasting longer than a softer camembert or brie. Many delicious cheeses that are soft as butter are made with unpasteurized "raw" milk, though in the United States, the FDA's concerns about these products' safety limit their sale.[8]

Yogurt is a fermented food with a long history and wide geographic distribution.[†] Local modifications of the techniques used to ferment milk have produced an array of liquid and solid yogurt products that vary in taste, texture, and tartness. Each uses added or native bacterial cultures to transform the milk sugar known as lactose to lactic acid, which produces the characteristic "tang" of the yogurt product.

† Fermentation is a metabolic process in which sugars are converted to organic acids, gases, or alcohol in the absence of molecular oxygen.

Part A: Homemade Ricotta Cheese

GOAL: To observe how adding acid breaks down the colloidal structure of milk into soft solid curds and liquid whey. The curds are collected and strained to become ricotta cheese.

EQUIPMENT

- 5-quart saucepan (medium stockpot)
- large spoon
- measuring cups, measuring spoons
- thermometer
- nonreactive bowl
- cutting board
- small knife
- lemon juicer
- cheesecloth
- fine-mesh strainer
- four 4-ounce plastic containers (your yield should be about two cups or 16 ounces)
- labeling tape

INGREDIENTS

- 8 cups milk (milk from sheep and goats can also be used with variable results)
- 1 cup heavy cream (satisfactory results are achieved without heavy cream, but the product will not have the richness that heavy cream provides)
- juice of one large lemon (usually about 3 tablespoons, but keep another lemon in reserve)
- pinch salt

Though it doesn't demonstrate the colloidal nature of milk, a vegan ricotta can be made from a firm tofu to which olive oil, lemon, miso paste, and seasonings have been added.[9]

INSTRUCTIONS

Figure 7.2 shows students making ricotta in our culinary laboratory.

1. **Prepare the strainer.** Line your large fine-mesh strainer with cheesecloth cut to overlap at least 4 inches on each side, and set it over the metal bowl. Set this on your work surface to have it ready for the curd-whey mixture.

2. **Lemon juice preparation.** Roll each lemon on the cutting board's surface to help release juices. Cut each lemon in half and use the juicer to extract the lemon juice. Remove the pits. Juice your second lemon to make sure you have at least 4 tablespoons.
3. **Heat to 93.3°C (200°F).** Place the milk, cream, and salt in the 5-quart saucepan and heat the mixture over medium-high heat, stirring to prevent scorching. Use your cooking thermometer to determine when the milk temperature reaches 93.3°C (200°F). At lower temperatures, the curds won't separate as effectively from the whey; at higher temperatures, some protein solids will begin to burn, and the texture will be ruined.
4. **Slowly add the lemon juice.**‡ Once the mixture has reached 93.3°C, turn off the burner and add the juice. Slowly pour the lemon juice into the heated milk and cream mixture while stirring constantly. The acid in the lemon juice will cause the curds to coagulate and separate from the whey. It may take 5 minutes or more for the separation to occur; be patient. You will see the opaque homogeneous milk begin to separate into the liquid whey and the soft solid curds, which float to the top of the liquid. Keep stirring until all the juice has been added. Add more juice if no curds are visible after 2–3 minutes.
5. **Let the mixture sit until it's thick.** Wait about 10 minutes for all the curds to separate from the whey. It's ready when the curds have floated to the top to form a thick layer, leaving the liquid whey underneath.
6. **Ladle the curds into the strainer.** Scoop out the thick top layer of curds and ladle them over the cheesecloth-covered strainer. Keep scooping out the curds until all that is left in the saucepan is the whey. You can discard the whey at this point.
7. **Let the ricotta drain.** Wait at least 10 minutes for the whey to completely drain from the ricotta through the cheesecloth into the bowl. Don't attempt to stir it or push it through the cheesecloth, as this will clog the holes in the cloth.
8. **Add a pinch of salt, and distribute the ricotta into containers.** Add cream to the desired thickness, and label with your group's name and date.

Part B—Fermentation of Milk Sugars to Make Yogurt

GOAL: To see that adding a live bacterial culture to the heated milk and keeping the mixture warm over several hours will facilitate fermentation that produces yogurt.

---

‡ Try substituting ¼ cup distilled white vinegar or citric acid for the lemon. For a more traditional touch, you can use animal rennet to separate the curds and whey. Mix 1 teaspoon of rennet with ¼ cup of cold water, then stir it into the milk mixture.

Figure 7.2 Steps in Making Ricotta

a. Four students in the gastronomy lab are happy to start making ricotta. Here they are heating the milk to 93.3°C, after which they will remove the warmed milk from the heat and add the lemon juice you can see on the bench.

b. Curds begin to separate from the whey after adding lemon juice (or other acid) to the heated milk. Patience.

c. Separate the curds from the whey using cheesecloth.

d. Curds can be stored with a pinch of salt and additional cream to adjust the consistency. These students were happy to pack some up to take home.

EQUIPMENT

- 1 Euro-Cuisine Yogurt Maker YMX650[§]
- 8 6-ounce glass yogurt jars
- 1 Waring Professional 2-ring burner
- 1 5-quart covered pot
- 1 Pyrex measuring cup
- 1 digital thermometer
- 1 large spoon
- 1 mixing bowl with pouring spout
- 1 potholder

INGREDIENTS

- 6 cups milk (plant-based milks such as coconut and almond milk can also be fermented)
- 6 ounces of yogurt with live cultures (starter cultures are available for plant-based milks)

INSTRUCTIONS

1. Procure your gallon of milk and starter yogurt from the front table.
2. Measure out 42 ounces of milk and transfer the milk to the 5-quart pot and put the pot on the induction burner set to medium-high heat. Stir constantly until the temperature of the liquid is 82.2°C (180°F), at which point it should begin to boil and creep up the sides of the pot (about 10 minutes). Boil at this temperature for 1–2 minutes with constant stirring.
3. Remove the pot from the heat and allow the contents to cool for a few minutes until it is safe to handle. Transfer the warm liquid to a mixing bowl.
4. Once the milk has been cooled to 43.3°C (110°F), add 6 ounces of yogurt starter. Adding the culture to milk that has not been cooled will kill the live cultures in the yogurt.
5. Stir vigorously to mix until there is an even consistency. Because the milk is opaque, it will be necessary to check the consistency by occasionally lifting the spoon and observing a stream of the mixture as you pour it back into the bulk (stirring 5 minutes).
6. Carefully transfer the mixture to the individual jars and place the jars WITHOUT TOPS in the Euro-Cuisine Yogurt Maker. Use the colored tape to identify your jars and lids. Assemble the yogurt maker with the

---

§ Though special commercially available yogurt incubators and jars can be used, they are not necessary. Just leaving the inoculated yogurt in a warm place overnight or for about twelve hours is usually enough to grow the bacteria and transform to liquid into a soft delicious solid.

filled jars in the lower chamber and the lid covers in the plastic upper chamber.

7. Set the timer for the number of hours that correlate with the fat content of your milk.

	with heating	without heating
raw milk	7 hours	8 hours
whole milk	7 hours	8 hours
2% milk	9 hours	10 hours

Note: Yogurt will be processed for the appropriate time and stored in the refrigerator until the next meeting.

8. Finally, and most importantly, clean and dry everything.

EXPERIMENT 4: SOUS VIDE—COOKING AT LOW PRESSURE WITHOUT OXYGEN

GOAL: To learn the sous vide cooking technique and to use it to prepare and present a sous vide main course dish with meat, fish, or vegetables as a culinary challenge.

Background

"Sous vide" is a French cooking term that is translated as "under a vacuum," a popular modernist cuisine cooking technique. A dish prepared using this method will be prepared in a vacuum pouch submerged in a circulating water bath at a carefully controlled temperature, often lower than usual cooking temperatures. The cooked foods can be held at this temperature longer without getting overdone.

What are the advantages of cooking in a vacuum? To answer that, we must first understand the disadvantages of cooking *in* the air. Remember that the air we breathe is a mixture of gases. About 78% of air is made up of nitrogen, an unreactive gas, which does not present a problem for cooking. The problem arises from the 20% of air that is the reactive combustible oxygen gas. We need oxygen: it is how we burn the sugars we eat and provide molecular energy in the form of ATP, but it can degrade foods while cooking.

Cooking without air protects foods from exposure to oxygen, which can degrade food quality and destroy flavors. Combining foods with oxygen before or during cooking can oxidize proteins, making them rubbery and interfering with other favorable Maillard reactions. When fats or oils combine with oxygen, they will taste rancid. Oxidation of sugars may interfere with caramelization reactions. Sous vide cooking results in even cooking and smooth, homogeneous textures. Figure 7.3 outlines the steps, equipment, and advantages of

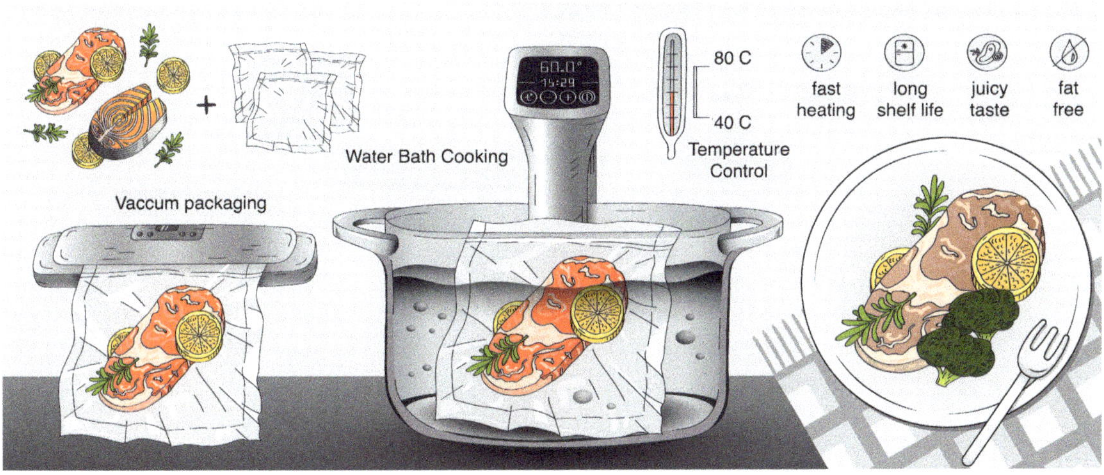

Figure 7.3 Sous Vide Cooking Technique

All the individual steps—food selection, sealing in airtight bags, submerging the food in a constant temperature water bath, and serving it—are shown in this infographic on sous side technology.

sous vide cooking. As you prepare your dishes, be sure to pay close attention to the changes in texture, color, and moisture content before and after cooking and as compared to your own experience of eating these foods cooked by conventional methods.

Cooking Meat Sous Vide

The ability of sous vide cooking to control the cooking temperature precisely and evenly can have amazing results on meat. In traditional cooking, heat travels by conduction from the outside to the inside, and even the tenderest cuts of meat can end up overcooked on the outside. This problem is avoided in sous vide cooking as the entire filet is heated to a constant temperature and never exceeds that temperature, creating an even level of doneness throughout the filet. With tougher cuts of meat, the two common ways of cooking, braising and roasting, both have the drawbacks of drying out the meat and overcooking it before it is tender enough to eat. Sous vide can bring something different to the table as well.

When cooking meat sous vide, you can avoid drying and overcooking the meat. By setting the temperature correctly, usually around 60°C (140°F), you can cook your brisket, pot roast, short ribs, or other "roast" cut to a perfect medium-rare temperature while still tenderizing it completely. Achieving these results is often a challenge with traditional cooking methods.

Cooking Seafood Sous Vide

One of the potential problems with cooking seafood over high heat is that it can become rubbery. Sous vide allows you to slowly cook the seafood over low heat, ensuring the result will be silky and tender. One of the struggles when cooking seafood with traditional methods is cooking to a precise temperature. Seafood quickly goes from undercooked to overcooked in the blink of an eye. It is nearly impossible to overcook food using sous vide. By setting the temperature between 52°C (rare) and 62°C (well done) (125°–144°F), you can be sure that your fish will cook to the exact doneness you are looking for. Salmon has a naturally delicious and unique flavor, so we recommend a light hand when adding additional sous vide ingredients to your pouch. Extra salt is not necessary.

Cooking Vegetables Sous Vide

A stumbling block when cooking many vegetables is trying to cook them all the way through without overcooking the surface. Sous vide allows you to achieve uniform doneness throughout vegetables and helps them hold together better. Should you decide to use the sous vide technique with your vegetables, cut them into pieces that are no thicker than 1/2 inch or so (~1 cm) and add only a small amount of liquid. Avoid crowding the vegetables in the bag. Salt, pepper, butter, and oil might be suitable additives.

Culinary Challenge

GOAL: To master the sous vide technique using a protein or vegetable of your choice and, by combining it with the unique ingredients from a food basket, create a main course to be evaluated by judges.

Sous Vide Beef Filet Medallion

Beef filet is cooked using the sous vide technique. Just before serving, the meat is quickly seared in a hot pan. The optimal result is even browning on all sides.

INGREDIENTS

- 2 beef filet medallions, about 2 inches thick, or other similar but more expensive beef cuts such as tenderloin or filet mignon
- Kosher salt and coarsely ground pepper, to taste
- 1 tablespoon (⅛ stick) unsalted butter
- 2 shallots, halved
- 2 sprigs fresh thyme

SOUS VIDE DIRECTIONS

1. Set the sous vide thermostat to 59°C (138°F) for medium-rare meat with the circulator on.
2. Season the beef filets with salt and pepper.
3. Place filets in separate small vacuum bags with 1 tablespoon butter, a shallot cut in half, and a thyme sprig in each bag.
4. Seal the bags to the desired vacuum; 90% to 95% vacuum is desirable for beef. LABEL THE BAGS WITH LABELING TAPE.
5. Once the bath has reached the target temperature of 59°C (138°F), place the bags in the circulating water bath.
6. Cook the beef to the desired doneness for about 60 minutes. You can hold the beef at this temperature for up to 90 minutes without affecting the quality or texture.

FINISHING DIRECTIONS (10 MINUTES BEFORE PLATING)

1. Remove the bag from the circulating water and remove the beef filet from the bag. Pat the filet with paper towels to dry and lightly season with salt and pepper. Note any color change of the beef after removal from the vacuum bag.
2. Warm the olive oil or butter in a hot pan. Add the filet, and quickly sear both sides.
3. Transfer the filets to a cutting board, and let rest for at least 3 minutes.
4. Cut into slices, garnish, and serve.

Sous Vide Salmon Filets

Salmon filets are cooked using the sous vide technique. Then, just before serving, the filets are quickly seared in a hot pan until browned, adding flavor and texture.

INGREDIENTS

- 2 6-ounce fresh coho salmon filets or other fish filets such as tilapia
- extra virgin olive oil (EVOO), butter, lemon wedges, or herbs (optional)

INSTRUCTIONS

1. Set the sous vide thermostat in the water bath to 52°C (125°F) with the circulator on.
2. Prepare a marinade for your salmon if you wish, but go lightly; salmon is a delicately flavored fish. The filets are from the ocean, so additional salt is not necessary. Add oil or butter if you wish, or add a squeeze of lemon or herbs. You may add nothing. You decide.

3. Put each filet into a cooking pouch and vacuum seal, being careful not to pull a vacuum that is so strong that the filets will deform. Once the plastic touches the flesh, hit the cancel button and seal. Label with colored tape.
4. Place the labeled pouches into the sous vide water bath and cook for at least 40 minutes. Cooking for too long might result in losing textural quality, but we will keep them warm at a lower temperature bath, so that won't be a problem.
5. Just before serving, remove the salmon. Pat them dry with paper towels.
6. If desired, heat the EVOO in a large skillet over high heat, watching carefully so the oil doesn't burn.
7. For crispy skin, pan sear in EVOO over high heat, place filets in the skillet, and quickly sear them to a golden brown on the skin side, about 1 minute at most. You may also choose not to crisp the skin, in which case you don't need to sear it.
8. Immediately serve with vegetables or starch. You may use any products available.

Sous Vide Asparagus[10]

Use whatever fresh vegetables are available and in season in your area.

INGREDIENTS

- 1 bunch asparagus (about 1 pound)
- 2 tablespoon butter
- salt and pepper (to taste)
- ¼ cup grated parmesan cheese
- 1 lemon (zested and juiced)
- salt and pepper (to taste)
- 2 cloves garlic (minced or grated)

INSTRUCTIONS

1. Set the sous vide thermostat to 82.2°C (180°F) with the circulator on.
2. Trim the woody ends of the asparagus.
3. Add the asparagus, butter, garlic, and a couple of heavy butter knives or spoons to a Food Saver bag and seal using a vacuum sealer. Alternatively, you can use a Ziploc bag and the water displacement method. (The butter knives help keep the bag weighted down in the water.)
4. Cook for 8–12 minutes based on the cooking recommendations below.

5. Remove from the water bath, season with salt and pepper, garnish with parmesan and lemon zest, and serve.

Cooking time will vary with the thickness of the asparagus according to the recommendations below.

Pencil thin asparagus: 8 minutes

Medium asparagus: 10 minutes

Thick asparagus: 12 minutes

COORDINATING PRESENTATION FOR SOUS VIDE CULINARY CHALLENGE For this experiment, you must prepare a full entrée for the judges that features your sous vide dish but adds complementary food tastes and textures using ingredients in your baskets. Given the restricted time of the lab, it may be necessary to prepare the sous vide ahead and plan how the other ingredients (usually starch and some additional vegetables) may be prepared while the protein is cooking. Plating is especially important here.

EXPERIMENT 5: VEGETABLE FOAMS

GOAL: You will experiment with stabilizers and emulsifiers to create two vegetable foams or éspumas, one made from potatoes and the other from winter squash.

Background

Foaming is one of the techniques most associated with modernist cooking. Foams are easy to make, very versatile, and fun to use and eat. Traditional cooking has used foams for a long time. These include whipped cream, beer heads, and soufflés. At the most basic level, foam is a structure that traps air in bubbles. Foams are like emulsions in this way. The continuous phase can be made from a variety of ingredients such as proteins (beer head), water (fruit mousse), or fat (whipped cream.) The texture of the foam is determined by the size of the bubbles and how much liquid is in the foam. Some foams are considered "set," meaning the structure has been solidified, such as when baking bread dough or a soufflé.

Today we will prepare two vegetable hot foams, éspumas, one made from potatoes (Yukon Gold potatoes have the right starch content) and the second made from a winter squash (buttercup or butternut squash works well).[11,12]

How Do Culinary Foams Work?

Foams are created by trapping air or gas bubbles in a liquid or solid substance, allowing both to become stable. Air is forced into the liquid using mechanical force from a blender or a mixer or by releasing compressed gas from a high-pressure capsule in a whipping siphon.

When creating the foam, the force provided to trap the bubbles must be greater than the natural tendency of the substance to break down to increase the surface area. Adding a stabilizer that will work as a surfactant is usually necessary to stabilize the foam. The stabilizer will coat the molecules and lessen the surface tension, allowing the molecules to adhere to each other more easily. Over time, without a surfactant, the bubble collapses, and the air is released and migrates to the surface because of differing densities.

Foams are also classified under colloids as they are composed of a mixture of a dispersed substance (usually air or other gas) in a continuous phase of solid or liquid particles.

What Do We Use for Stabilizers or Emulsifiers?

Thicker initial substances will result in denser foams, but some general guidelines for the amount of stabilizer to use are as follows:

Agar—0.25% to 1.0%

Iota carrageenan—0.2% to 1.0%

Gelatin—0.4% to 1.7%

Methylcellulose—1.0% to 3.0%

Xanthan gum—0.1% to 0.4%

Other thickeners can be used, such as egg whites, lecithin, and aquafaba (made from chickpea juice for a vegan foam). Different stabilizers can change the textures and taste of the foams, so you may want to experiment.

Foaming Equipment

Many tools can be used to create foams, and each tool creates a slightly different texture. All the tools aim to introduce air into the liquid you are foaming. The three tools used in this lab, the whipping siphon, the tamis, and the scraper, are described below. A list of other tools commonly used to make foams is also included. For devices such as whisks and immersion blenders, you want to make sure part of the tool is out of the liquid so it will carry air into the foam.[13]

Whipping Siphon

The whipping siphon is an excellent tool for making foams of all kinds. It is a container you fill with liquid and then pressurize with small cartridges of compressed gas such as nitrous oxide (N_2O) or, occasionally, nitrogen (N_2) or carbon dioxide (CO_2). Siphons are very effective at creating foams and help in the storage of mixtures. N_2O will be forced into the emulsion to aerate it and is preferred due to its increased solubility in the continuous phase. N_2O is harmless, though it is used by dentists as "laughing gas" to provide patients undergoing minor dental surgery with an altered mental state characterized by euphoria and peace. Detailed instructions on using the whipping siphon can be found online.[14]

Tamis

A tamis is a fine-screened sieve or food mill that helps remove lumps and vegetable fibers from the cooked fruit and vegetables to create velvety smooth purees to serve as the base for the foams. It is shaped like a snare drum with cylindrical sides (usually metal) and with a fine metal mesh across the bottom. Food to be processed is placed in the middle of the tamis and pushed through with a scraper.

Scraper

Special scrapers are used to push the food through the fine mesh of the tamis. The strength and flexibility of silicon make it a favorite choice for this kitchen utensil.

Other Kitchen Equipment Used to Make Foams

WHISKS, MANUAL AND ELECTRIC Whisks are a great way to create foams whose texture can vary from light to dense. Manual whisks can get the job done, but using an electric whisk attachment dramatically speeds up the process and tends to make finer foams. Whisk attachments can be immersion blenders or mixers.

STANDING OR HANDHELD MIXER Mixers without a whisk attachment can also create lighter foams very efficiently.

MILK FROTHER A milk frother is an inexpensive tool used to create foam for cappuccinos or lattes. When used with modernist ingredients, it can create similar foams from other liquids.

IMMERSION BLENDER Immersion blenders are good at creating airs and other light foams. Ensuring part of the blade is out of the liquid is crucial. A traditional standing blender will not work well for foaming because the blades are completely submerged.

AQUARIUM BUBBLER An aquarium pump or bubbler is one of the more unusual ways to create foams. It works well for creating large bubbles, similar to soap bubbles.

Foams Using a Siphon: Éspumas

Éspumas is the Spanish word for "foam" or "froth."

A culinary foam consists of natural flavors like fruit juices or vegetable purees, soup, and stock bases mixed with gelling or stabilizing agents such as lecithin, gelatin, or natural fats in cream and other dairy products. Air is introduced using either a mechanical technique of whipping the fluids with a handheld immersion blender or extruding through a cream whipper using nitrous oxide cartridges.[15]

This technique has been used since the 1970s to make dairy foams such as cappuccinos and whipped cream. Using this technique and equipment, Ferran Adrià took the culinary world by storm in the 1990s by expanding foaming to include fruits, vegetables, and other ingredients. The benefit of this technique is that when you incorporate air mechanically, in a fierce manner, into a very intense and robust flavored sauce, you expand the flavor, so it becomes light and sumptuous, and the volume doubles.

A wispy foam is created with a handheld immersion blender, while a dense mousse-type foam is created with a cream whipper or siphon to create an éspumas.

ÉSPUMAS AND FOAMING FACTS

- The liquid or puree must be thick or dense enough to hold its shape.
- For the foam to hold its shape, some thickening or gelling agent must be present in the liquid.
- Thickening and gelling agents are gelatin, lecithin, agar, and natural fats such as butter, cream, and other dairy products.
- For hot foams, the best thickeners are fat or starch; this can be found in butter, cream, or milk. It is essential to ensure the liquid is not too hot. The perfect temperature is between 50°C and 65°C. Place the cream whipper in a pot filled with hot water; do not keep it in the pot for longer than 2 hours.
- There are many different gases available to charge the cream whipper. Cartridges containing either nitrogen (N_2) or nitrous oxide (N_2O) are most commonly used for vegetable foams. Carbon dioxide (CO_2) is used to make carbonated drinks or foams.

Parts A and B—General Instructions

Because we do not have enough whipping siphons for each group to have two, Part A starts by splitting the established groups into two teams. The Potato Team has half of the students (at least one from each group) who must collaborate to prepare two siphons of potato éspuma. The Squash Team is made up of the other half of the students (again, at least one per group) who collaborate to prepare two siphons filled with squash éspuma. Students then return to their groups to complete the culinary challenge and construct and plate two dishes, savory and sweet, using the squash éspuma and potato éspuma they have prepared.

Part A: Preparation of Potato Éspuma by the Potato Team

Preparation of the potato puree: approximately 1,000 grams of potato (Yukon Gold or another kind) are peeled, thinly sliced, and simmered in salted water with 1 teaspoon of lemon juice for approximately 20 minutes or until tender.

Since the Potato Team must prepare two 1 liters of whipping siphons, each filled with approximately 250 grams of warm potato puree, it is best to split the group in half, each preparing one siphon. Each 1 liter siphon will be enough to share between two groups for the culinary challenge.

INGREDIENTS FOR 1 LITER WHIPPING SIPHON

- 250 grams warm potato puree
- 150 grams vegetable broth, cold
- 1.25 grams iota carrageenan
- 1 gram xanthan gum
- 125 grams heavy cream (vegan and nondairy options available)
- 35 grams butter (vegan butter substitute available)

INSTRUCTIONS

1. Disperse the carrageenan and xanthan gum into the vegetable broth. It will appear lumpy at this point. Heat the mixture while stirring to 95°C (203°F) and hold for 3 minutes to hydrate fully. Use the timer to keep track of the time. Remove from heat.
2. Add the butter to the warm broth, and stir until entirely melted. Add the heavy cream slowly to the buttered broth, making sure not to curdle the cream by adding it to a liquid that is too hot.
3. Finally, fold in the warmed pureed potatoes and season with salt. Check for consistency of the éspumas; it should be thicker than soup but thin enough to be poured. Adjust with warm vegetable broth if the mixture is too thick, or add a bit more puree if it is too thin.
4. Follow the directions to assemble your whipping siphon. Transfer all your éspumas to your whipping siphon and charge with one or two nitrous oxide cartridges.
5. Hold in a 60°C (140°F) bath until ready to dispense but not longer than 1 hour. Be sure to shake the whipper six times before dispensing a ½ cup at a time. If the foam is a bit runny, shake a few more times.
6. Your culinary challenge using your potato foam is detailed below.

Part B: Preparation of the Squash Éspumas by the Squash Team

Approximately 1,000 grams buttercup squash (or other dense, deeply colored fall squash) must be peeled, seeded, and processed into 2-inch cubes, then simmered in boiling salted water until tender.

Since the Squash Team must prepare two 1 liter whipping siphons filled with approximately 250 grams of warm squash puree, it is best to split the group in half, each preparing one siphon. Each 1 liter siphon will be enough to share between two groups for the culinary challenge.

INGREDIENTS FOR 1 LITER WHIPPING SIPHON

- 250 grams warm squash puree
- 150 grams vegetable broth, cold
- 1 gram iota carrageenan
- 1 gram xanthan gum
- 125 grams heavy cream (vegan and nondairy substitutes are available)
- 35 grams butter (vegan butter substitute is available)
- season to taste

INSTRUCTIONS

1. Disperse the carrageenan and xanthan gum into the cold broth. Heat the mixture to 95°C (203°F) with constant stirring and hold for 3 minutes to hydrate fully. Use a timer to keep track of the time. Remove from heat.
2. Add the butter to the hot broth and stir until the butter is melted. Add the heavy cream slowly, ensuring the temperature is not high enough to curdle the cream.
3. Gently fold in warm, pureed squash, a bit at a time, and season with salt.
4. Check for consistency of the éspumas. It should be thicker than soup but thin enough to pour. Adjust with warm vegetable broth if it is too thick. If it is too thin, check with the teaching assistant or instructor.
5. Follow the directions to assemble your whipping siphon. Transfer all the éspumas to a whipping siphon and charge with nitrous oxide cartridges.
6. Hold in a 60°C (140°F) bath until ready to dispense but not longer than 1 hour. Be sure to shake the whipper six times before dispensing a ½ cup at a time. If the foam is a bit runny, shake a few more times.
7. Your culinary challenge using your potato foam is detailed below.

Part C: Culinary Challenge

Each team will choose a basket from the front of the room and use its unique ingredients and the shared ingredients on the central table to create two vegetable foam éspumas: one made from potatoes and the other from squash.

Dishes should be formally plated and presented to the judges for tasting. You will have 30 minutes to complete this challenge.

SHARED SUPPLIES Staple ingredients, such as olive oil, canola oil, avocado oil, balsamic vinegar, white wine vinegar, rice vinegar, half and half, butter, onions, garlic, maple syrup, ginger, honey, sugar, brown sugar, lemon, and finishing salts, will be made available. Small bowls, spoons, and other edible small bowls can be made available for presentation.

BASKET SUPPLIES Four baskets are prepared and randomly chosen by the groups. Each basket will contain at least three garnishes to make a sweet éspuma and three garnishes to make a savory éspuma. Savory garnishes might include leeks, scallions, shallots, garlic scapes, carrots, peppers, beets, spinach, microgreens, mint, thyme, or wasabi. Sweet garnishes might include marshmallows, graham crackers, gingersnaps, toffee, pomegranate seeds, fresh cranberries, citrus, pepita seeds, sesame seeds (watch for allergies), or granola.

EXPERIMENT 6: SPHERIFICATION

GOAL: In this experiment, you will learn about the chemistry involved with the spherification of liquids, learn the pros and cons of the reverse spherification technique, and apply this technique to make a mocktail and strawberry amuse-bouche.

Background

Spherification is a great trick and a quintessentially modernist process that transforms a liquid into an orb enveloped by a gelled skin, resulting in beads that resemble caviar. There is a certain childlike joy in biting into these tiny spheres and getting a burst of flavor that saturates your palate with fresh, flavorful ingredients.

The secret to spherification is forming a gel on a droplet's surface and not extending it to the interior. To do this, we use sodium alginate and calcium chloride, which, when combined, form a gel in the presence of water. The sodium alginate is mixed in with the fruit juice and cannot be set on its own. The calcium ions needed to trigger the gelling process are in the setting bath, so as the mixture is dropped, little orbs of flavor form instantaneously.

Removing the spheres from the calcium bath yields liquid droplets surrounded by a tissue-thin gel skin. Be careful; leaving them in the calcium bath too long transforms them into chewy blobs.[16] Sodium alginate is an extract of brown seaweed. This reagent is used to thicken pie fillings and improve the texture of ice cream.

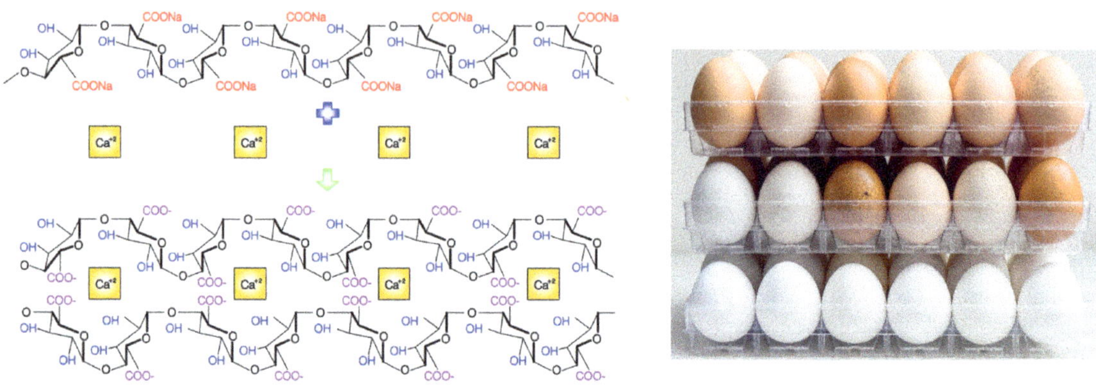

$$2NaC_6H_7O_6 + Ca\,Cl_2 \rightarrow 2NaCl + C_{12}\,H_{14}\,CaO_{12}$$

Figure 7.4 Sodium Alginate Precipitation with Calcium

Alginate is a derivatized polysaccharide that is soluble when its carboxylate groups are bound to sodium (Na^+). On the addition of calcium (Ca^{+2}), the sodium is displaced, and the Ca-alginate forms an insoluble film. An analogy is how a plastic egg carton (alginate) surrounds an egg (calcium).

- Sodium alginate is also used to form the "cherries" in canned cherry pie filling.
- Sodium citrate is a salt of citric acid. It promotes gelling by binding to stray ions. Sodium citrate is also an emulsifier.
- Calcium chloride is a mineral salt of calcium. Calcium chloride reacts with sodium alginate to form a solid gel in water.
- Xanthan gum is created by bacterial fermentation. It takes only a minuscule amount of xanthan gum to thicken the liquid, making it popular with modernist thickeners.

What Is Happening Chemically?

Alginate is derived from a carbohydrate, and initially it is complexed with either sodium (Na^+) or hydrogen (H^+) ions. With these +1 ions, the alginate is soluble and flexible. However, when mixed with a +2 ion, such as calcium (Ca^{+2}), the +2 ion displaces the +1 ion. The structure becomes more rigid, and cross-links form, creating a shell around the alginate liquid.

Basic Spherification

In basic spherification, the soluble sodium form of the alginate is mixed with the liquid to be spherified, and the solution is dropped carefully into a calcium bath. As the droplet enters the bath, the calcium displaces the sodium, and the calcium alginate forms on the surface, creating the sphere. Pearl-like droplets of uniform shape and size are created. Because of their similar appearance to fish eggs, the spheres are often called caviar.

Reverse Spherification

The reverse spherification technique consists of reversing the solution that has the alginate. This technique adds calcium to the liquid to be spherified and dissolves the alginate in the bath. Submerging a liquid with calcium in the alginate bath produces a sphere that can be a lot larger. If the juice or flavored liquid to be spherified does not naturally contain calcium, calcium lactate or calcium lactate gluconate is added.

PROS OF REVERSE SPHERIFICATION

- Reverse spherification is more versatile than basic spherification as it can make spheres with almost any product. It can be used with liquids with a high calcium or alcohol content, which makes them great for cocktails and dairy products like cheese, milk, and yogurt.
- The resulting sphere is long-lasting and can be stored for later consumption. Unlike basic spherification, gelification stops when the sphere is removed from the sodium alginate bath and rinsed with water, allowing you to prepare them beforehand.
- Reverse spherification results in a sphere with a thicker membrane than with basic spherification. Thanks to this, the resulting spheres can be manipulated easily, they conserve their shape better when plated (spheres produced with basic spherification flatten and acquire an orb or egg yolk shape when plated), and they can be used in more ways (such as fillings in sponge cakes or mousses).
- Gelification can still occur when the liquid to be spherified is acidic because the gelling process in reverse spherification occurs on the surface of the sphere. The sodium alginate in the bath does not penetrate the globe, and a translucent layer of gel is created around the main ingredient. By contrast, in basic spherification, the gelling process occurs internally and has the color of the juice.
- The main ingredient's consistency and flavor are not altered by adding calcium lactate gluconate and calcium lactate as they have no discernible flavor and dissolve in liquid without changing their density. We do not use calcium chloride in reverse spherification because it is very salty.

CONS OF REVERSE SPHERIFICATION

- The thicker membrane of these spheres is more evident on your palate. You still get the "pop" sensation, but you can detect the solid jelly in addition to the liquid contents.
- The sodium alginate bath must rest for twelve to twenty-four hours to eliminate the air bubbles created by dissolving the sodium alginate with the immersion blender.
- The flavored liquid may need to be thickened with xanthan gum, and if air bubbles get trapped in the process, you may need to wait a few hours to eliminate them.
- Getting a perfect sphere on the plate with reverse spherification is more challenging than with basic spherification. When you pour the main ingredient into the viscous bath, the spheres will stick together if you aren't careful.

Part A: Basic Spherification to Make Mocktails

INGREDIENTS

A soda such as ginger ale or flavored seltzer can be paired with any non-acidic natural fruit spheres to create the mocktail. This recipe can use spheres made from grape, mango, cherry, apple, carrot, blueberry, and acai berry or coconut water.

FOR THE FRUIT SPHERES

- 180 grams fruit juice (180 milliliter—¾ cup—6 ounces)
- 15 grams sugar (1.5 tablespoons)
- 5.0 grams lemon juice (5 milliliters—1 teaspoon)
- 2.5 grams sodium alginate—(1 teaspoon)
- 0.3 gram xanthan gum—(pinch or about ⅛ teaspoon)
- 1.5 grams sodium citrate (¼ teaspoon)

FOR THE CALCIUM BATH

- 500 grams of filtered distilled water (500 milliliters or 2 ⅛ cup)
- 2.5 grams calcium chloride (½ teaspoon)

INSTRUCTIONS

STEPS 1–5 MUST BE DONE 6 TO 24 HOURS IN ADVANCE!

1. Combine juice, sugar, and lemon juice in the clear bowl associated with the immersion blender. Mix until sugar is dissolved.
2. Mix sodium alginate, xanthan gum, and sodium citrate in a small disposable beaker. Sprinkle the powdery mixture over the surface and allow the suspension to rest for 5 minutes. The floating powders should begin aggregating, and the surface should "crack" to reveal juice below.
3. Blend the mixture until smooth using an immersion blender. To reduce the number of bubbles formed, keep the blade below the surface of the juice. Once completely blended, the mixture should have the consistency of thick cream. Xanthan gum can be added if it is too thin, and juice can be added if it is too thick.
4. Pass the mixture through a fine sieve or de-gas using vacuum aspiration so that air bubbles will dissipate. The presence of bubbles will result in an irregular sphere shape. Refrigerate for at least 2 hours.

STEPS 5–13 CAN BE DONE IN THE LAB.

5. Blend filtered water and calcium chloride with a whisk until the powder fully dissolves.

6. Prepare three small bowls of filtered water to rinse spheres in step 10.
7. Fill a syringe (or turkey baster) with the juice mixture and position the needle about 6 inches above the bath's surface. If the syringe is too high, the droplets will flatten on hitting the water surface. The droplets will form tails like raindrops if the syringe is too low.
8. Gently dispense the gel base, one drop at a time, directly into the calcium chloride bath. Move the syringe around to form a single layer.
9. Allow the spheres to rest in the calcium chloride bath for at least 2–4 minutes to form a strong membrane. The sodium alginate reacts with the calcium chloride to form a membrane during this time. The longer the spheres sit in the bath, the thicker the membrane becomes, so don't wait too long. A slotted spoon can be used to transfer the spheres from the calcium chloride bath to the first water bath.
10. Continue rinsing the spheres by transferring them to the second water bath and then to the third water bath. Triple rinsing the spheres in clean water halts the gelling reaction and eliminates any bitter flavor left over from the calcium chloride bath.
11. Repeat steps 7–10 with the remaining juice mixture or until you have as many little spheres as you want.
12. Seltzer, ginger ale, and clear fruit juice will be available. Martini glasses are a great way to display your mocktail.
13. See Part C for the culinary challenge.

Part B: Reverse Spherification—Strawberry or Mango Amuse-Bouche[17]

INGREDIENTS

FOR THE FRUIT PUREE

- 250 grams (8.8 ounces) strawberry puree made from (~300 grams / 10.6 ounces strawberries)
- 10 grams (0.35 ounce) sugar
- pinch of salt
- 5 grams calcium lactate gluconate (2%)
- Strawberry juice to store the spheres if you are not planning to consume them within a few hours.

FOR THE ALGINATE BATH

- 1,000 grams (35 ounces) filtered distilled water
- 5 grams sodium alginate (0.5%)

Figure 7.5 Student Preparations of Mocktails and Strawberry Globes

INSTRUCTIONS

PREPARING THE STRAWBERRY OR MANGO PUREE THE NIGHT BEFORE

1. Puree the strawberries or mango using an immersion blender.
2. Pass the fruit puree through a tamis or fine sieve.
3. Add the sugar, salt, and calcium lactate gluconate. Blend.
4. Cover in plastic wrap and store it in the refrigerator overnight to remove the air bubbles.

PREPARING THE SPHERIFICATION BATH THE NIGHT BEFORE

5. Dispersing and hydrating sodium alginate. Mix the sodium alginate with the distilled water using a blender until completely dissolved. If this is the first time you are doing this, be aware that it may take longer than expected.
6. Removing air bubbles. Strain the mixture and store it in the fridge covered in plastic wrap overnight to eliminate the air bubbles.

STEPS 7–12 CAN BE DONE ON THE DAY OF LAB—CREATING FRUIT GLOBES WITH REVERSE SPHERIFICATION

7. Gather the strawberry or mango puree with calcium content from the fridge, the sodium alginate bath, and a large spoon to make large spheres.
8. Prepare another bowl with plain water that you will use later to rinse the spheres to remove the excess sodium alginate.
9. Fill the spoon with the strawberry puree. Wipe the bottom with a paper towel, place the spoon over the bath, slightly touching its surface, and ease the puree into the sodium alginate bath as carefully as possible. Use your spoon to separate the spheres so that the shells harden uniformly. This technique takes practice. Experiment with factors like how fast to lower the puree into the bath, the tools you use, and so on. Experiment, and see what works!
10. Stir the bath gently with the slotted spoon without touching the spheres. If you let them sit in the bottom of the bath, they will flatten; if you allow them to float, the top won't be covered with the sodium alginate solution and won't gel. The trick is to gently bathe the top surface of the sphere with liquid from the bath. Make sure the spheres don't touch each other, or they will stick. Start with one sphere at a time until you get used to the process. A flat-bottomed pan may be better than a bowl if you plan to make multiple spheres.
11. Wait for at least 5 minutes. The longer you wait, the thicker the gel that will form. For an ideal outcome, you want the gel layer surrounding the sphere to be as thin as possible, but it must also be strong enough to hold the shape and allow for careful handling. If the membrane is too fragile, it may easily break when you remove the spheres from the bath or place them on the serving spoon. If the membrane is too thick, reduce the time.
12. Carefully remove the spheres one at a time from the sodium alginate bath using a slotted spoon and rinse them in the bowl with clean distilled water.

Tips: Always start with one globe first to adjust the pouring process and the time in the sodium alginate bath. If the sphere membrane is too thin and easily breaks when handling it carefully with the slotted spoon or when plating it, extend the time in the calcium bath until you get the desired strength. Remember that the thinner the membrane, the better experience people will have when eating it.

Part C: Culinary Challenge

Each group will choose a basket from the front table and, using these unique garnishes, prepare a mocktail and amuse-bouche for each judge. Remember to add the spheres to the mocktails just before serving. Enjoy eating and drinking the rest.

Notes

CHAPTER 1

1. Colin Towell, *Essential Survival Skills* (London: Penguin, 2011), https://www.dk.com/us/book/9780756659981-essential-survival-skills/.
2. U.S. Congress. Nutrition Labeling and Education Act of 1990. Pub. L. No. 535, U.S. Statutes at Large, vol. 104, 101st Cong., 2nd sess. (1990). https://www.govinfo.gov/app/details/STATUTE-104/STATUTE-104-Pg2353.
3. W. O. Atwater, *Foods: Nutritive Value and Cost* (Washington, DC: U.S. Dept. of Agriculture, 1894), http://archive.org/details/CAT87201446.
4. Janet A. Novotny, Sarah K. Gebauer, and David J. Baer, "Discrepancy between the Atwater Factor Predicted and Empirically Measured Energy Values of Almonds in Human Diets," *American Journal of Clinical Nutrition* 96, no. 2 (August 2012): 296–301, https://doi.org/10.3945/ajcn.112.035782.
5. National Cancer Institute, "Definition of Homeostasis—NCI Dictionary of Cancer Terms—NCI," nciAppModulePage, February 2, 2011, nciglobal, ncienterprise, https://www.cancer.gov/publications/dictionaries/cancer-terms/def/homeostasis.
6. Mark A. Febbraio and Michael Karin, "Sweet Death: Fructose as a Metabolic Toxin That Targets the Gut-Liver Axis," *Cell Metabolism* 33, no. 12 (December 2021): 2316–28, https://doi.org/10.1016/j.cmet.2021.09.004.
7. *Sugar: The Bitter Truth* (2009), https://www.youtube.com/watch?v = dBnniua6-oM.
8. D. M. Klurfeld et al., "Lack of Evidence for High Fructose Corn Syrup as the Cause of the Obesity Epidemic," *International Journal of Obesity* 37, no. 6 (June 2013): 771–73, https://doi.org/10.1038/ijo.2012.157.
9. World Health Organization (WHO), "Obesity and Overweight," https://www.who.int/news-room/fact-sheets/detail/obesity-and-overweight.
10. David S. Ludwig et al., "The Carbohydrate-Insulin Model: A Physiological Perspective on the Obesity Pandemic," *American Journal of Clinical Nutrition* 114, no. 6 (December 2021): 1873–85, https://doi.org/10.1093/ajcn/nqab270.
11. David S. Ludwig and Cara B. Ebbeling, "The Carbohydrate-Insulin Model of Obesity: Beyond 'Calories In, Calories Out,'" *JAMA Internal Medicine* 178, no. 8 (August 1, 2018): 1098, https://doi.org/10.1001/jamainternmed.2018.2933.
12. Neira Sáinz et al., "Leptin Signaling as a Therapeutic Target of Obesity," *Expert Opinion on Therapeutic Targets* 19, no. 7 (July 3, 2015): 893–909, https://doi.org/10.1517/14728222.2015.1018824.
13. Cleveland Clinic, "9 Vitamins and Minerals You Should Take Daily," November 4, 2021, https://health.clevelandclinic.org/which-vitamins-should-you-take/.

CHAPTER 2

1. "Taste and Smell," BrainFacts.org, April 1, 2010, https://www.brainfacts.org:443/thinking-sensing-and-behaving/taste/2012/taste-and-smell.
2. Nirupa Chaudhari and Stephen D. Roper, "The Cell Biology of Taste," *Journal of Cell Biology* 190, no. 3 (August 9, 2010): 285–96, https://doi.org/10.1083/jcb.201003144.
3. Jessleen K. Kanwal, "Brain Tricks to Make Food Taste Sweeter: How to Transform Taste Perception and Why It Matters," *Science in the News* (blog), January 11, 2016, https://sitn.hms.harvard.edu/flash/2016/brain-tricks-to-make-food-taste-sweeter-how-to-transform-taste-perception-and-why-it-matters/.
4. Linda Buck and Richard Axel, "A Novel Multigene Family May Encode Odorant Receptors: A Molecular Basis for Odor Recognition," *Cell* 65, no. 1 (April 1991): 175–87, https://doi.org/10.1016/0092–8674(91)90418-X.
5. Eric H. Holbrook and Donald A. Leopold, "An Updated Review of Clinical Olfaction," *Current Opinion in Otolaryngology & Head & Neck Surgery* 14, no. 1 (February 2006): 23–28, https://doi.org/10.1097/01.moo.0000193174.77321.39.
6. Crispian Scully, *Oral and Maxillofacial Medicine: The Basis of Diagnosis and Treatment*, 3rd ed. (Edinburgh: Churchill Livingstone/Elsevier, 2013).
7. Stephen A. Gravina, Gregory L. Yep, and Mehmood Khan, "Human Biology of Taste," *Annals of Saudi Medicine* 33, no. 3 (May 2013): 217–22, https://doi.org/10.5144/0256–4947.2013.217.
8. Joanna Jeruzal-Świątecka, Wojciech Fendler, and Wioletta Pietruszewska, "Clinical Role of Extraoral Bitter Taste Receptors," *International Journal of Molecular Sciences* 21, no. 14 (July 21, 2020): 5156, https://doi.org/10.3390/ijms21145156.
9. A. L. Fox, "The Relationship between Chemical Constitution and Taste," *Proceedings of the National Academy of Sciences of the United States of America* 18, no. 1 (January 1932): 115–20, https://doi.org/10.1073/pnas.18.1.115.
10. Stephen Wooding, "Phenylthiocarbamide: A 75-Year Adventure in Genetics and Natural Selection," *Genetics* 172, no. 4 (April 1, 2006): 2015–23, https://doi.org/10.1093/genetics/172.4.2015.
11. G. E. Dubois et al., "Concentration-Response Relationships of Sweeteners: A Systematic Study," in *Sweeteners: Discovery, Molecular Design and Chemoreception*, ed. D. E. Walters, F. T. Orthoefer, and G. E. Dubois. Vol. 450. ACS Symposium Series. Washington, DC: American Chemical Society, 1991, https://doi.org/10.1021/bk-1991–0450.
12. Stephen D. Roper, "TRPs in Taste and Chemesthesis," in *Mammalian Transient Receptor Potential (TRP) Cation Channels*, ed. Bernd Nilius and Veit Flockerzi, vol. 223, *Handbook of*

Experimental Pharmacology (Cham: Springer, 2014), 827–71, https://doi.org/10.1007/978-3-319–05161–1_5.
13. Alissa A. Nolden, Gabrielle Lenart, and John E. Hayes, "Putting out the Fire—Efficacy of Common Beverages in Reducing Oral Burn from Capsaicin," *Physiology & Behavior* 208 (September 2019): 112557, https://doi.org/10.1016/j.physbeh.2019.05.018.
14. "Atomic Fireball," Ferrara Candy Shop, accessed July 1, 2023, https://www.ferraracandyshopusa.com/.
15. Alissa A. Nolden and John E. Hayes, "Perceptual and Affective Responses to Sampled Capsaicin Differ by Reported Intake," *Food Quality and Preference* 55 (January 2017): 26–34, https://doi.org/10.1016/j.foodqual.2016.08.003.
16. Kartik Venkatachalam and Craig Montell, "TRP Channels," *Annual Review of Biochemistry* 76, no. 1 (June 7, 2007): 387–417, https://doi.org/10.1146/annurev.biochem.75.103004.142819.
17. David Julius, "TRP Channels and Pain," *Annual Review of Cell and Developmental Biology* 29, no. 1 (October 6, 2013): 355–84, https://doi.org/10.1146/annurev-cellbio-101011–155833.
18. Nao Kokaji and Masashi Nakatani, "With a Hint of Sudachi: Food Plating Can Facilitate the Fondness of Food," *Frontiers in Psychology* 12 (October 15, 2021): 699218, https://doi.org/10.3389/fpsyg.2021.699218.
19. Amy Fleming, "How Sound Affects the Taste of Our Food," *The Guardian*, March 11, 2014, sec. Life and Style, https://www.theguardian.com/lifeandstyle/wordofmouth/2014/mar/11/sound-affects-taste-food-sweet-bitter.
20. Massimiliano Zampini and Charles Spence, "The Role of Auditory Cues in Modulating the Perceived Crispness and Staleness of Potato Chips," *Journal of Sensory Studies* 19, no. 5 (October 2004): 347–63, https://doi.org/10.1111/j.1745–459x.2004.080403.x.
21. David Julius and Ardem Patapoutian, "Discoveries of Receptors for Temperature and Touch," Nobel Prize in Physiology or Medicine 2021, https://www.nobelprize.org/prizes/medicine/2021/advanced-information/.
22. Yalda Moayedi, Lucia F. Duenas-Bianchi, and Ellen A. Lumpkin, "Somatosensory Innervation of the Oral Mucosa of Adult and Aging Mice," *Scientific Reports* 8, no. 1 (July 2, 2018): 9975, https://doi.org/10.1038/s41598–018–28195–2.
23. Dengyong Liu et al., "Impact of Oral Processing on Texture Attributes and Taste Perception," *Journal of Food Science and Technology* 54, no. 8 (July 2017): 2585–93, https://doi.org/10.1007/s13197–017–2661–1.
24. Stevens Institute of Technology, "Direct Touch of Food Makes Eating Experience More Enjoyable," *ScienceDaily*, February 5, 2020, https://www.sciencedaily.com/releases/2020/02/200205151511.htm.
25. Małgorzata Starowicz, "Analysis of Volatiles in Food Products," *Separations* 8, no. 9 (September 17, 2021): 157, https://doi.org/10.3390/separations8090157.
26. James Briscione and Brooke Parkhurst, *The Flavor Matrix: The Art and Science of Pairing Common Ingredients to Create Extraordinary Dishes* (New York: Houghton Mifflin Harcourt, 2018).
27. "Foodpairing®—Better Products, Faster," July 24, 2021, https://www.foodpairing.com/.

CHAPTER 3

1. Alvi Imdad, "Difference between Conduction, Convection and Radiation in Tabular Form," *Ox Science* (blog), August 14, 2023, https://oxscience.com/conductionconvection-and-radiation/.
2. "What Is Heat Transfer? Conduction, Convection, Radiation and FAQs," BYJUS, https://byjus.com/physics/heat-transfer-conduction-convection-and-radiation/.
3. "Microwave Ovens," http://hyperphysics.phy-astr.gsu.edu/hbase/Waves/mwoven.html.
4. Center for Devices and Radiological Health, "Microwave Ovens," FDA, October 12, 2023, https://www.fda.gov/radiation-emitting-products/resources-you-radiation-emitting-products/microwave-ovens.
5. Jenny Dorsey, "Infrared Cooking 101," Institute of Culinary Education, January 22, 2021, https://www.ice.edu/blog/infrared-cooking.
6. Chris Woodford, "How Do Induction Cooktops Work?," Explain that Stuff, April 8, 2011, http://www.explainthatstuff.com/induction-cooktops.html.
7. "Induction Stoves and EMF Testing," California Energy Commission, September 25, 2020, https://efiling.energy.ca.gov/GetDocument.aspx?tn=234914&DocumentContentId=67776.
8. Melissa Clark, "The Case for Induction Cooking, Versus Gas Stoves," *New York Times*, March 11, 2022, sec. Food, https://www.nytimes.com/2022/03/11/dining/induction-cooking.html.
9. "Malted Milk," Wikipedia, September 9, 2023, https://en.wikipedia.org/w/index.php?title = Malted_milk&oldid = 1174605959.
10. Diogo N. De Oliveira, Maico De Menezes, and Rodrigo R. Catharino, "Thermal Degradation of Sucralose: A Combination of Analytical Methods to Determine Stability and Chlorinated Byproducts," *Scientific Reports* 5, no. 1 (April 15, 2015): 9598, https://doi.org/10.1038/srep09598.
11. American Cancer Society, "Aspartame and Cancer Risk," accessed July 1, 2023, https://www.cancer.org/cancer/risk-prevention/chemicals/aspartame.html.
12. Miryam Naddaf, "Aspartame Is a Possible Carcinogen: The Science behind the Decision," *Nature*, July 14, 2023, d41586–023–02306–0, https://doi.org/10.1038/d41586–023–02306–0.
13. Smithsonian Magazine and K. Annabelle Smith, "Why the Tomato Was Feared in Europe for More Than 200 Years," *Smithsonian Magazine*, accessed July 1, 2023, https://www.smithsonianmag.com/arts-culture/why-the-tomato-was-feared-in-europe-for-more-than-200-years-863735/.
14. Kirstie Canene-Adams et al., "The Tomato as a Functional Food," *Journal of Nutrition* 135, no. 5 (May 2005): 1226–30, https://doi.org/10.1093/jn/135.5.1226.
15. Lyndsay Cordell, "The Reason Wagyu Beef Prices Are More Expensive Than Angus," April 4, 2022, https://www.wideopencountry.com/wagyu-beef-price/.
16. Heather Bates, "Why Is Toro So Expensive?," WAGYUMAN, October 1, 2020, https://wagyuman.com/blogs/resources/why-is-toro-so-expensive.
17. Harold McGee, *On Food and Cooking: The Science and Lore of the Kitchen*, Completely rev. and updated (New York: Scribner, 2004).
18. Lisa Bramen, "Around the World in 80 Eggs," *Smithsonian Magazine*, May 12, 2010, https://www.smithsonianmag.com/arts-culture/around-the-world-in-80-eggs-87987638/?no-ist.
19. Condé Nast, "Why Are There 100 Folds in a Chef's Toque?," Bon Appétit, April 1, 2012, https://www.bonappetit.com/people/chefs/article/why-are-there-100-folds-in-a-chef-s-toque.
20. "Egg-Yolk101–187x300–1.Jpg (290×428)," Egg Nutrition Facts, accessed July 1, 2023, https://eggs.ab.ca/wp-content/uploads/2021/03/Egg-Yolk101–187x300–1.jpg.
21. Sudhir C. Roy, "Low-Density Lipoproteins from Egg Yolk: A Natural Carrier of Highly Emulsifying Species," *New Food*

Magazine, February 28, 2013, https://www.newfoodmagazine.com/article/10090/low-density-lipoproteins-from-egg-yolk-a-natural-carrier-of-highly-emulsifying-species/.

22. Meram Chalamaiah et al., "Physicochemical and Functional Properties of Livetins Fraction from Hen Egg Yolk," *Food Bioscience* 18 (June 2017): 38–45, https://doi.org/10.1016/j.fbio.2017.04.002.
23. Chen Sun and Shicui Zhang, "Immune-Relevant and Antioxidant Activities of Vitellogenin and Yolk Proteins in Fish," *Nutrients* 7, no. 10 (October 22, 2015): 8818–29, https://doi.org/10.3390/nu7105432.
24. Vida Šimat et al., "Astaxanthin from Crustaceans and Their Byproducts: A Bioactive Metabolite Candidate for Therapeutic Application," *Marine Drugs* 20, no. 3 (March 12, 2022): 206, https://doi.org/10.3390/md20030206.
25. Grant Alkin, "Ceviche—Science and Food," March 26, 2013, https://www.scienceandfood.org/ceviche/.
26. "Tmp4697_thumb1.Jpg (965×768)," What-When-How.com, http://what-when-how.com/wp-content/uploads/2011/05/tmp4697_thumb1.jpg.
27. Harold McGee, "Caramelization: New Science, New Possibilities," Curious Cook, accessed July 1, 2023, https://www.curiouscook.com/site/2012/09/caramelization-new-science-new-possibilities.html.
28. Sandman, "The Science behind Caramelization," *The Food Untold* (blog), August 5, 2021, https://thefooduntold.com/food-chemistry/the-science-behind-caramelization/.
29. J. E. Hodge, "Dehydrated Foods, Chemistry of Browning Reactions in Model Systems," *Journal of Agricultural and Food Chemistry* 1, no. 15 (October 1953): 928–43, https://doi.org/10.1021/jf60015a004.
30. Sarah Everts, "The Maillard Reaction Turns 100," *Chemical & Engineering News*, October 1, 2012, https://cen.acs.org/articles/90/i40/Maillard-Reaction-Turns-100.html.
31. Nahid Tamanna and Niaz Mahmood, "Food Processing and Maillard Reaction Products: Effect on Human Health and Nutrition," *International Journal of Food Science* (2015): 1–6, https://doi.org/10.1155/2015/526762.

CHAPTER 4

1. "The Science of Ceviche—Is It Cooked?," SBS Food, accessed July 1, 2023, https://www.sbs.com.au/food/article/the-science-of-ceviche-is-it-cooked/jwi7c9ydh.
2. Douglas Rodriguez and Laura Zimmerman, *The Great Ceviche Book* (Berkeley, CA: Ten Speed Press, 2003).
3. Bjørn Tore Lunestad et al., "Microbiota of Lutefisk, a Nordic Traditional Cod Dish with a High pH," *Food Control* 90 (August 1, 2018): 312–16, https://doi.org/10.1016/j.foodcont.2018.03.011.
4. Harold McGee, "For Old-Fashioned Flavor, Bake the Baking Soda," *New York Times*, September 14, 2010, sec. Food, https://www.nytimes.com/2010/09/15/dining/15curious.html.
5. Alice Popovici, "The Colonel in the Kitchen: A Surprising History of Sous Vide," NPR, May 9, 2018, sec. Food History & Culture, https://www.npr.org/sections/thesalt/2018/05/09/608308624/the-colonel-in-the-kitchen-a-surprising-history-of-sous-vide.
6. Popovici, "The Colonel in the Kitchen."
7. Chun Z. Yang et al., "Most Plastic Products Release Estrogenic Chemicals: A Potential Health Problem That Can Be Solved," *Environmental Health Perspectives* 119, no. 7 (July 1, 2011): 989–96, https://doi.org/10.1289/ehp.1003220.
8. Center for Food Safety and Applied Nutrition, "Bisphenol A (BPA): Use in Food Contact Application," FDA, October 19, 2023, https://www.fda.gov/food/food-packaging-other-substances-come-contact-food-information-consumers/bisphenol-bpa-use-food-contact-application.
9. Lisa Zimmermann et al., "Benchmarking the In Vitro Toxicity and Chemical Composition of Plastic Consumer Products," *Environmental Science & Technology* 53, no. 19 (October 1, 2019): 11467–77, https://doi.org/10.1021/acs.est.9b02293.
10. "Denis Papin," *Encyclopedia Britannica*, accessed August 18, 2024, https://www.britannica.com/biography/Denis-Papin.
11. Tim Chin, "How Pressure Cookers Actually Work," Serious Eats, May 7, 2023, https://www.seriouseats.com/how-pressure-cookers-work.
12. U.S. Department of Agriculture, "High Altitude Cooking," June 13, 2015, http://www.fsis.usda.gov/food-safety/safe-food-handling-and-preparation/food-safety-basics/high-altitude-cooking.
13. Sergi Maicas, "The Role of Yeasts in Fermentation Processes," *Microorganisms* 8, no. 8 (July 28, 2020): 1142, https://doi.org/10.3390/microorganisms8081142.
14. Patrick McGovern et al., "Early Neolithic Wine of Georgia in the South Caucasus," *Proceedings of the National Academy of Sciences* 114, no. 48 (November 28, 2017), https://doi.org/10.1073/pnas.1714728114.
15. "Speyer Wine Bottle," Wikipedia, accessed August 21, 2023, https://en.wikipedia.org/w/index.php?title = Speyer_wine_bottle&oldid = 1171570600.
16. A. Goffeau et al., "Life with 6000 Genes," *Science* 274, no. 5287 (October 25, 1996): 546–67, https://doi.org/10.1126/science.274.5287.546.

CHAPTER 5

1. Stevens Institute of Technology, "Direct Touch of Food Makes Eating Experience More Enjoyable," ScienceDaily, February 5, 2020, https://www.sciencedaily.com/releases/2020/02/200205151511.htm.
2. Melissa Clark, "Mayonnaise: Oil, Egg and a Drop of Magic," *New York Times*, May 22, 2012, sec. Food, https://www.nytimes.com/2012/05/23/dining/easy-homemade-mayonnaise.html.
3. Jacob Burton, "What Is an Emulsion & How Does it Work?," YouTube video, 9:48, October 12, 2011, https://www.youtube.com/watch?v=u2JSiyolnwo.
4. Madju, "What Is the Difference between Stabilizers and Emulsifiers," December 6, 2022, https://www.differencebetween.com/what-is-the-difference-between-stabilizers-and-emulsifiers/.
5. Sabrina J. Hickenbottom et al., "Structure of a Lipid Droplet Protein," *Structure* 12, no. 7 (July 2004): 1199–1207, https://doi.org/10.1016/j.str.2004.04.021.
6. Seyed Hossein Davoodi et al., "Health-Related Aspects of Milk Proteins," *Iranian Journal of Pharmaceutical Research: IJPR* 15, no. 3 (2016): 573–91.
7. Center for Food Safety and Applied Nutrition, "Raw Milk," FDA, October 24, 2023, https://www.fda.gov/food/resources-you-food/raw-milk.
8. Center for Food Safety and Applied Nutrition, "Raw Milk Misconceptions and the Danger of Raw Milk Consumption," FDA, October 24, 2023, https://www.fda.gov/food/buy-store-serve-safe-food/raw-milk-misconceptions-and-danger-raw-milk-consumption.
9. Emad A. Abada, "Application of Microbial Enzymes in the Dairy Industry," in *Enzymes in Food Biotechnology*, ed. Mohammed Kuddus (Amsterdam: Elsevier, 2019), 61–72, https://doi.org/10.1016/B978–0-12–813280–7.00005–0.

10. Breanna, "Breeds of Dairy Cows," *Whatcom Family Farmers* (blog), accessed July 1, 2023, https://whatcomfamilyfarmers.org/breeds-of-dairy-cows/.
11. David Julian McClements, Emily Newman, and Isobelle Farrell McClements, "Plant-Based Milks: A Review of the Science Underpinning Their Design, Fabrication, and Performance," *Comprehensive Reviews in Food Science and Food Safety* 18, no. 6 (November 2019): 2047–67, https://doi.org/10.1111/1541–4337.12505.
12. "U.S. Population: Most Eaten Brands of Whipped Topping (Cream Type) 2011–2020," Statista, accessed July 1, 2023, https://www.statista.com/statistics/281278/us-households-most-eaten-brands-of-whipped-topping-cream-type-trend/.
13. Center for Food Safety and Applied Nutrition, "FDA's Nutrition Initiatives," FDA, October 19, 2023, https://www.fda.gov/food/food-labeling-nutrition/fdas-nutrition-initiatives.
14. David Julian McClements and Lutz Grossmann, "A Brief Review of the Science behind the Design of Healthy and Sustainable Plant-Based Foods," *Npj Science of Food* 5, no. 1 (June 3, 2021): 17, https://doi.org/10.1038/s41538–021–00099-y.
15. Xiaoyi Jiang et al., "Lactic Acid Bacteria as Structural Building Blocks in Non-Fat Whipping Cream Analogues," *Food Hydrocolloids* 135 (February 1, 2023): 108137, https://doi.org/10.1016/j.foodhyd.2022.108137.

CHAPTER 6

1. United Nations, "Day of 8 Billion," accessed July 1, 2023, https://www.un.org/en/dayof8billion.
2. Population Connection, "Population Milestones," accessed July 1, 2023, https://populationconnection.org/learn/population-milestones/.
3. United Nations, "What Is Climate Change?," accessed July 1, 2023, https://www.un.org/en/climatechange/what-is-climate-change.
4. United Nations, "The Paris Agreement," accessed July 1, 2023, https://www.un.org/en/climatechange/paris-agreement.
5. Melissa Denchak, "Paris Climate Agreement: Everything You Need to Know," Natural Resources Defense Council, February 19, 2021, https://www.nrdc.org/stories/paris-climate-agreement-everything-you-need-know.
6. Action Against Hunger, "Ending World Hunger & Malnutrition," accessed July 1, 2023, https://www.actionagainsthunger.org/.
7. Kirsten Jaglo, Shannon Kenny, and Jenny Stephenson, "From Farm to Kitchen: The Environmental Impacts of U.S. Food Waste," U.S. Environmental Protection Agency Office of Research and Development, November 2021, https://www.epa.gov/system/files/documents/2021–11/from-farm-to-kitchen-the-environmental-impacts-of-u.s.-food-waste_508-tagged.pdf.
8. U.S. Environmental Protection Agency (EPA), "Wasted Food Scale," Overviews and Factsheets, August 12, 2015, https://www.epa.gov/sustainable-management-food/wasted-food-scale.
9. Jean Buzby, "Food Waste and Its Links to Greenhouse Gases and Climate Change," U.S. Department of Agriculture, January 24, 2022, https://www.usda.gov/media/blog/2022/01/24/food-waste-and-its-links-greenhouse-gases-and-climate-change.
10. Manny Fox Morone, "Forget Expiration Dates: Spoilage Sensors Could Tell Us When Food Actually Goes Bad," *Chemical & Engineering News*, May 17, 2020, https://cen.acs.org/analytical-chemistry/diagnostics/Forget-expiration-dates-Spoilage-sensors/98/i19.
11. Project Drawdown, "Drawdown Foundations," February 5, 2020, https://drawdown.org/drawdown-foundations.
12. J. Poore and T. Nemecek, "Reducing Food's Environmental Impacts through Producers and Consumers," *Science* 360, no. 6392 (June 2018): 987–92, https://doi.org/10.1126/science.aaq0216.
13. Julia Moskin et al., "Your Questions about Food and Climate Change, Answered," *New York Times*, April 30, 2019, sec. Food, https://www.nytimes.com/interactive/2022/dining/climate-change-food-eating-habits.html.
14. Nadja Popovich, Cari Vander Yacht, and Eden Weingart, "Quiz: How Does Your Diet Contribute to Climate Change?," *New York Times*, April 30, 2019, sec. Climate, https://www.nytimes.com/interactive/2019/04/30/climate/your-diet-quiz-global-warming.html.
15. Niko Kommenda et al., "This Is the Environmental Cost of the Food We Eat," *Washington Post*, February 1, 2023, https://www.washingtonpost.com/climate-environment/interactive/2023/food-impact-climate-water-wildlife/.
16. Brian Vines, "Ugly Food Fight: Misfits Market, Imperfect Foods, and the Battle against Food Waste," *Consumer Reports*, January 12, 2022, https://www.consumerreports.org/health/food-shopping/ugly-food-fight-misfits-market-imperfect-foods-food-waste-a6326488257/.
17. Misfits Market, "Grocery Delivery FAQs," accessed July 1, 2023, https://www.misfitsmarket.com/faq.
18. CMI Orchards, "I'm Perfect," accessed July 1, 2023, https://www.cmiapples.com/catalog/brands/im-perfect.
19. Maria Godoy, "Wal-Mart, America's Largest Grocer, Is Now Selling Ugly Fruit and Vegetables," NPR, July 20, 2016, sec. Producers, https://www.npr.org/sections/thesalt/2016/07/20/486664266/walmart-world-s-largest-grocer-is-now-selling-ugly-fruit-and-veg.
20. Marie Yeung, "Microbial Forensics in Food Safety," *Microbiology Spectrum* 4, no. 4 (August 12, 2016): 10.1128/microbiolspec.emf-0002–2013, https://doi.org/10.1128/microbiolspec.emf-0002–2013.
21. Rep. Frank Pallone, "H.R.2901—118th Congress (2023–2024): Food Labeling Modernization Act of 2023," Pub. L. No. 2901, 118th Congress (2023), https://www.congress.gov/bill/118th-congress/house-bill/2901.
22. Federico Bottari and Cecilia Mark-Herbert, "Development of Uniform Food Information—the Case of Front of Package Nutrition Labels in the EU," *Archives of Public Health* 80, no. 1 (December 2022): 175, https://doi.org/10.1186/s13690–022–00915–1.
23. Apeel, "What's in Apeel?," August 29, 2022, https://www.apeel.com/blog/whats-in-apeel.
24. XiaoZhi Lim, "Xiaogang Liu Is Using Fluorescence to Tackle the Problem of Illicit Cooking Oil," *Chemical & Engineering News*, May 2, 2020, https://cen.acs.org/food/food-science/Xiaogang-Liu-using-fluorescence-tackle/98/i17.
25. "AgTech Post-Harvest Innovator That Helps Extend Produce Quality," AgroFresh, accessed July 1, 2023, https://www.agrofresh.com/.
26. Ryp Labs, "Fighting Food Waste," accessed July 1, 2023, https://www.ryplabs.com.
27. Gottfried J. Palm et al., "Structure of the Plastic: Degrading Ideonella Sakaiensis MHETase Bound to a Substrate," *Nature Communications* 10, no. 1 (April 12, 2019): 1717, https://doi.org/10.1038/s41467–019–09326–3.

28. Alex Scott, "Nestlé Will Spend $2 Billion to Peel Away from Virgin Plastic," *Chemical & Engineering News*, January 23, 2020, https://cen.acs.org/materials/polymers/Nestlspend-2-billion-peel-away/98/i4.
29. "Ice Cream Cone," Wikipedia, November 25, 2023, https://en.wikipedia.org/w/index.php?title=Ice_cream_cone&oldid=1186794616.
30. Prachi Patel, "The Time Is Now for Edible Packaging," *Chemical & Engineering News*, January 26, 2020, https://cen.acs.org/food/food-science/time-edible-packaging/98/i4.
31. U.S. Food and Drug Administration (FDA), "Food Code 2022," November 6, 2023, https://www.fda.gov/food/fda-food-code/food-code-2022.
32. White House, "Biden-Harris Administration National Strategy on Hunger, Nutrition, and Health," September 2022, https://www.whitehouse.gov/wp-content/uploads/2022/09/White-House-National-Strategy-on-Hunger-Nutrition-and-Health-FINAL.pdf.
33. "One Generation Away," *OneGenAway* (blog), August 11, 2022, https://onegenaway.com/.
34. World Central Kitchen, "World Central Kitchen," accessed July 1, 2023, https://wck.org/.
35. José Andrés, "Today, One Year after the Brutal Invasion of Ukraine Started," *Twitter*, February 24, 2023, https://twitter.com/chefjoseandres/status/1629224004660899840.
36. World Central Kitchen, "WCK Serves 12+ Million Meals in the Month since Türkiye-Syria Quakes," March 10, 2023, https://wck.org/news/turkiye-syria-update2.
37. José Andrés, "We Need a Secretary of Food," *New York Times*, December 8, 2020, sec. Opinion, https://www.nytimes.com/2020/12/08/opinion/covid-pandemic-food-crisis.html.
38. Cornell University Department of Food Science, "Biomaterials and Biointerfaces," The *Goddard Research Group* (blog), accessed July 1, 2023, https://blogs.cornell.edu/goddard/.
39. Stephanie M. Andler and Julie M. Goddard, "Transforming Food Waste: How Immobilized Enzymes Can Valorize Waste Streams into Revenue Streams," *Npj Science of Food* 2, no. 1 (October 29, 2018): 19, https://doi.org/10.1038/s41538–018–0028–2.
40. Viviane P. Romani, Vilásia G. Martins, and Julie M. Goddard, "Radical Scavenging Polyethylene Films as Antioxidant Active Packaging Materials," *Food Control* 109 (March 2020): 106946, https://doi.org/10.1016/j.foodcont.2019.106946.
41. Kevin P. Sullivan et al., "Mixed Plastics Waste Valorization through Tandem Chemical Oxidation and Biological Funneling," *Science* 378, no. 6616 (October 14, 2022): 207–11, https://doi.org/10.1126/science.abo4626.
42. Hélène Deveau, Josiane E. Garneau, and Sylvain Moineau, "CRISPR/Cas System and Its Role in Phage-Bacteria Interactions," *Annual Review of Microbiology* 64, no. 1 (October 13, 2010): 475–93, https://doi.org/10.1146/annurev.micro.112408.134123.
43. "The Nobel Prize in Chemistry 2020," NobelPrize.org, accessed July 1, 2023, https://www.nobelprize.org/prizes/chemistry/2020/summary/.
44. Syed Shan-e-Ali Zaidi et al., "New Plant Breeding Technologies for Food Security," *Science* 363, no. 6434 (March 29, 2019): 1390–91, https://doi.org/10.1126/science.aav6316.
45. National Institutes of Health (NIH), "Gene Editing—Digital Media Kit," March 7, 2019, https://www.nih.gov/news-events/gene-editing-digital-press-kit.
46. Ewen Callaway, "CRISPR Plants Now Subject to Tough GM Laws in European Union," *Nature* 560, no. 7716 (August 2018): 16, https://doi.org/10.1038/d41586–018–05814–6.
47. Corteva Agriscience, "Welcome to Corteva Agriscience," accessed July 1, 2023, https://www.corteva.com.
48. Abby Meyer and Sara Dastgheib-Vinarov, "The Future of Food? CRISPR-Edited Agriculture," *Food and Drug Law Institute (FDLI)* (blog), November 19, 2021, https://www.fdli.org/2021/11/the-future-of-food-crispr-edited-agriculture/.
49. Inari, "Seeding Change to Bring Value across the Food System to Unlock Radical Possibilities," accessed July 1, 2023, https://inari.com/.
50. TreeCo, "R&D Engine of the Forest Industry," accessed July 1, 2023, https://tree-co.com/.
51. Shaobo Wei et al., "A Transcriptional Regulator That Boosts Grain Yields and Shortens the Growth Duration of Rice," *Science* 377, no. 6604 (July 22, 2022): eabi8455, https://doi.org/10.1126/science.abi8455.
52. Miguel Lima et al., "A Narrative Review of Alternative Protein Sources: Highlights on Meat, Fish, Egg and Dairy Analogues," *Foods* 11, no. 14 (July 11, 2022): 2053, https://doi.org/10.3390/foods11142053.
53. Alexandra Katidi et al., "Nutritional Quality of Plant-Based Meat and Dairy Imitation Products and Comparison with Animal-Based Counterparts," *Nutrients* 15, no. 2 (January 12, 2023): 401, https://doi.org/10.3390/nu15020401.
54. Mark R. O'Brian, "How Does the Impossible Burger Look and Taste Like Real Beef?," *Discover Magazine*, April 22, 2019, https://www.discovermagazine.com/planet-earth/how-does-the-impossible-burger-look-and-taste-like-real-beef.
55. Xin Guan et al., "Bioprocessing Technology of Muscle Stem Cells: Implications for Cultured Meat," *Trends in Biotechnology* 40, no. 6 (June 2022): 721–34, https://doi.org/10.1016/j.tibtech.2021.11.004.
56. "Cultured Meat," Wikipedia, December 6, 2023, https://en.wikipedia.org/w/index.php?title=Cultured_meat&oldid=1188670576.
57. Good Food Institute, "The Science of Cultivated Meat," January 27, 2021, https://gfi.org/science/the-science-of-cultivated-meat/.
58. "The Jetsons," Wikipedia, December 8, 2023, https://en.wikipedia.org/w/index.php?title=The_Jetsons&oldid=1188884367.
59. Ruban Whenish et al., "A Framework for the Sustainability Implications of 3D Bioprinting through Nature-Inspired Materials and Structures," *Bio-Design and Manufacturing* 5, no. 2 (April 2022): 412–23, https://doi.org/10.1007/s42242–021–00168-x.
60. byFlow, "3D Shaping the Future of Food . . . Today," accessed July 1, 2023, https://www.3dbyflow.com/.
61. Jonathan David Blutinger et al., "Precision Cooking for Printed Foods via Multiwavelength Lasers," *Npj Science of Food* 5, no. 1 (September 1, 2021): 24, https://doi.org/10.1038/s41538–021–00107–1.
62. Róisín Burke, ed., *Handbook of Molecular Gastronomy: Scientific Foundations and Culinary Applications* (Boca Raton, FL: CRC Press, 2021).
63. Stephen Ashley, "Synthetic Food: Better Cooking through Chemistry," PBS, June 5, 2013, https://www.pbs.org/wgbh/nova/article/synthetic-food-better-cooking-through-chemistry/.

PART IV

1. "ServSafe—Get Certified at Home," National Restaurant Association Educational Foundation, accessed July 1, 2023, http://servsafe.com/Home.

CHAPTER 7

1. "Science of Eggs: Anatomy of an Egg," Exploratorium: Museum of Science, Art and Human Perception, accessed July 1, 2023, https://annex.exploratorium.edu/cooking/eggs/eggcomposition.html.
2. La Cocina de Gisele, "Can You Buy 100 Year Old Eggs?," accessed July 1, 2023, https://lacocinadegisele.com/knowledgebase/can-you-buy-100-year-old-eggs.
3. MasterClass, "Century Eggs Explained: How to Make Century Eggs—2023," December 4, 2023, https://www.masterclass.com/articles/century-eggs-explained-plus-century-egg-recipe.
4. Williams Sonoma, "Rösle Egg Topper," accessed July 1, 2023, https://www.williams-sonoma.com/products/rosle-egg-topper/.
5. W. Wayt Gibbs and Nathan Myhrvold, "Cryogenic Cooking," *Scientific American* 305, no. 2 (August 2011): 31.
6. JuJu Kim, "Woman Loses Stomach after Drinking Liquid Nitrogen Cocktail," *Time*, October 9, 2012, https://newsfeed.time.com/2012/10/09/woman-loses-stomach-after-drinking-liquid-nitrogen-cocktail/.
7. Dakin Farm, "Maple Sugar on Snow," accessed July 1, 2023, https://www.dakinfarm.com/recipe.aspx/confection/Maple-Sugar-on-Snow.
8. U.S. Department of Agriculture (USDA), "Is It Safe to Eat Cheese Made from Unpasteurized Milk?," March 6, 2023, https://ask.usda.gov/s/article/Is-it-safe-to-eat-cheese-made-from-unpasteurized-milk.
9. Jasmine Briones, "Easy Tofu Ricotta Cheese (Vegan + Just 7-Ingredients)," Sweet Simple Vegan, June 21, 2023, https://sweetsimplevegan.com/tofu-ricotta-cheese/.
10. Erin, "How to Sous Vide Asparagus (with 3 Flavoring Ideas!)," Platings + Pairings, September 2022, https://www.platingsandpairings.com/sous-vide-asparagus/#recipe.
11. Madalene, "How to Make Culinary Foams, Air and Espumas," British Larder, April 2, 2015, https://www.britishlarder.co.uk/how-to-make-culinary-foams-air-and-espumas/.
12. Helen Rennie, "Random Beets and How to Make Vegetable Foams," *Beyond Salmon* (blog), July 29, 2012, http://www.beyondsalmon.com/2012/07/random-beets-and-how-to-make-vegetable.html.
13. James Logsdon, *Modernist Cooking Made Easy: Getting Started* (CreateSpace Independent Publishing Platform, 2012), https://www.goodreads.com/book/show/17085591-modernist-cooking-made-easy.
14. *iSi Gourmet Whip How-to-Use*. YouTube video, 2015. https://www.youtube.com/watch?v=uK2T9WxjUvc.
15. Kitchen Theory, "Lecithin & Culinary Foam," *Immersive Dining* (blog), accessed February 15, 2023, https://kitchen-theory.com/lecithin-culinary-foam/.
16. Nathan Myhrvold, *Modernist Cuisine*, vol. 4: *The Art and Science of Cooking* (Seattle: Cooking Lab, 2011), 184–95.
17. "Strawberry Spheres with Ground Pepper," Molecular Recipes, December 12, 2013, http://www.molecularrecipes.com/spherification-1/strawberry-spheres-ground-pepper/.

Bibliography

Abada, Emad A. "Application of Microbial Enzymes in the Dairy Industry." In *Enzymes in Food Biotechnology*, ed. Mohammed Kuddus, 61–72. Amsterdam: Elsevier, 2019. https://doi.org/10.1016/B978–0-12–813280–7.00005–0.

Action Against Hunger. "Ending World Hunger & Malnutrition." Accessed July 1, 2023. https://www.actionagainsthunger.org/.

AgroFresh. "AgTech Post-Harvest Innovator That Helps Extend Produce Quality." Accessed December 11, 2023. https://www.agrofresh.com/.

Alkin, Grant. "Ceviche—Science and Food." March 26, 2013. https://www.scienceandfood.org/ceviche/.

American Cancer Society, "Aspartame and Cancer Risk." Accessed July 1, 2023. https://www.cancer.org/cancer/risk-prevention/chemicals/aspartame.html.

Andler, Stephanie M., and Julie M. Goddard. "Transforming Food Waste: How Immobilized Enzymes Can Valorize Waste Streams into Revenue Streams." *Npj Science of Food* 2, no. 1 (October 29, 2018): 19. https://doi.org/10.1038/s41538–018–0028–2.

Andrés, José. "Today, One Year after the Brutal Invasion of Ukraine Started." Twitter, February 24, 2023. https://twitter.com/chefjoseandres/status/1629224004660899840.

———. "We Need a Secretary of Food." *New York Times*, December 8, 2020, sec. Opinion. https://www.nytimes.com/2020/12/08/opinion/covid-pandemic-food-crisis.html.

Apeel. "What's in Apeel?" August 29, 2022. https://www.apeel.com/blog/whats-in-apeel.

Ashley, Stephen. "Synthetic Food: Better Cooking through Chemistry." PBS, June 5, 2013. https://www.pbs.org/wgbh/nova/article/synthetic-food-better-cooking-through-chemistry/.

"Atomic Fireball." Ferrara Candy Shop. Accessed July 1, 2023. https://www.ferraracandyshopusa.com/.

Atwater, W. O. *Foods: Nutritive Value and Cost.* Washington, DC: U.S. Department of Agriculture, 1894. http://archive.org/details/CAT87201446.

Ball, Aimee Lee. "Hervé This and the Future of Food." *New York Times*, September 17, 2015, sec. *T Magazine*. https://www.nytimes.com/2015/09/17/t-magazine/herve-this-nbn-future-food.html.

Bates, Heather. "Why Is Toro So Expensive?" Wagyuman, October 1, 2020. https://wagyuman.com/blogs/resources/why-is-toro-so-expensive.

Blutinger, Jonathan David, Alissa Tsai, Erika Storvick, Gabriel Seymour, Elise Liu, Noà Samarelli, Shravan Karthik, Yorán Meijers, and Hod Lipson. "Precision Cooking for Printed Foods via Multiwavelength Lasers." *Npj Science of Food* 5, no. 1 (September 1, 2021): 24. https://doi.org/10.1038/s41538–021–00107–1.

Bottari, Federico, and Cecilia Mark-Herbert. "Development of Uniform Food Information—the Case of Front of Package Nutrition Labels in the EU." *Archives of Public Health* 80, no. 1 (December 2022): 175. https://doi.org/10.1186/s13690–022–00915–1.

BrainFacts.org. "Taste and Smell." April 1, 2010. https://www.brainfacts.org:443/thinking-sensing-and-behaving/taste/2012/taste-and-smell.

Bramen, Lisa. "Around the World in 80 Eggs." *Smithsonian Magazine*, May 12, 2010. https://www.smithsonianmag.com/arts-culture/around-the-world-in-80-eggs-87987638/?no-ist.

Brar, Abheetinder. "Colloids." Chemistry LibreTexts, October 2, 2013. https://chem.libretexts.org/Bookshelves/Physical_and_Theoretical_Chemistry_Textbook_Maps/Supplemental_Modules_(Physical_and_Theoretical_Chemistry)/Physical_Properties_of_Matter/Solutions_and_Mixtures/Colloid.

Breanna. "Breeds of Dairy Cows." *Whatcom Family Farmers* (blog). Accessed July 1, 2023. https://whatcomfamilyfarmers.org/breeds-of-dairy-cows/.

Briones, Jasmine. "Easy Tofu Ricotta Cheese (Vegan + Just 7-Ingredients)." Sweet Simple Vegan, June 21, 2023. https://sweetsimplevegan.com/tofu-ricotta-cheese/.

Briscione, James, and Brooke Parkhurst. *The Flavor Matrix: The Art and Science of Pairing Common Ingredients to Create Extraordinary Dishes.* New York: Houghton Mifflin Harcourt, 2018.

Buck, Linda, and Richard Axel. "A Novel Multigene Family May Encode Odorant Receptors: A Molecular Basis for Odor Recognition." *Cell* 65, no. 1 (April 1991): 175–87. https://doi.org/10.1016/0092–8674(91)90418-X.

Burke, Róisín M., Alan L. Kelly, Christophe Lavelle, and Hervé This, and Vo Kientza, eds. *Handbook of Molecular Gastronomy: Scientific Foundations, Educational Practices, and Culinary Applications.* Boca Raton, FL: CRC Press, 2021. https://doi.org/10.1201/9780429168703.

Burton, Jacob. *What Is An Emulsion and How Does it Work?* YouTube video, 9:48, October 12, 2011. https://www.youtube.com/watch?v=u2JSiyolnwo.

Buzby, Jean. "Food Waste and Its Links to Greenhouse Gases and Climate Change." U.S. Department of Agriculture, January 24, 2022. https://www.usda.gov/media/blog/2022/01/24/food-waste-and-its-links-greenhouse-gases-and-climate-change.

byFlow. "3D Shaping the Future of Food . . . Today." Accessed July 1, 2023. https://www.3dbyflow.com/.

BYJUS. "What Is Heat Transfer? Conduction, Convection, Radiation and FAQs." Accessed July 1, 2023. https://byjus.com/physics/heat-transfer-conduction-convection-and-radiation/.

Callaway, Ewen. "CRISPR Plants Now Subject to Tough GM Laws in European Union." *Nature* 560, no. 7716 (August 2018): 16. https://doi.org/10.1038/d41586-018-05814-6.

Canene-Adams, Kirstie, Jessica K. Campbell, Susan Zaripheh, Elizabeth H. Jeffery, and John W. Erdman. "The Tomato as a Functional Food." *Journal of Nutrition* 135, no. 5 (May 2005): 1226–30. https://doi.org/10.1093/jn/135.5.1226.

Center for Devices and Radiological Health. "Microwave Ovens." October 12, 2023. https://www.fda.gov/radiation-emitting-products/resources-you-radiation-emitting-products/microwave-ovens.

Center for Food Safety and Applied Nutrition. "Bisphenol A (BPA): Use in Food Contact Application." FDA, October 19, 2023. https://www.fda.gov/food/food-packaging-other-substances-come-contact-food-information-consumers/bisphenol-bpa-use-food-contact-application.

———. "FDA's Nutrition Initiatives." October 19, 2023. https://www.fda.gov/food/food-labeling-nutrition/fdas-nutrition-initiatives.

———. "Raw Milk." October 24, 2023. https://www.fda.gov/food/resources-you-food/raw-milk.

———. "Raw Milk Misconceptions and the Danger of Raw Milk Consumption." October 24, 2023, https://www.fda.gov/food/buy-store-serve-safe-food/raw-milk-misconceptions-and-danger-raw-milk-consumption.

"Century Egg." Wikipedia. Accessed September 26, 2024. https://en.wikipedia.org/wiki/Century_egg.

Chalamaiah, Meram, Yussef Esparza, Feral Temelli, and Jianping Wu. "Physicochemical and Functional Properties of Livetins Fraction from Hen Egg Yolk." *Food Bioscience* 18 (June 2017): 38–45. https://doi.org/10.1016/j.fbio.2017.04.002.

Chaudhari, Nirupa, and Stephen D. Roper. "The Cell Biology of Taste." *Journal of Cell Biology* 190, no. 3 (August 9, 2010): 285–96. https://doi.org/10.1083/jcb.201003144.

Chin, Tim. "How Pressure Cookers Actually Work." Serious Eats, May 7, 2023. https://www.seriouseats.com/how-pressure-cookers-work.

Clancy, L. J., J. A. Critchley, A. G. Leitch, B. J. Kirby, A. Ungar, and D. C. Flenley. "Arterial Catecholamines in Hypoxic Exercise in Man." *Clinical Science and Molecular Medicine* 49, no. 5 (November 1975): 503–6. https://doi.org/10.1042/cs0490503.

Clark, Melissa. "The Case for Induction Cooking, versus Gas Stoves." *New York Times*, March 11, 2022, sec. Food. https://www.nytimes.com/2022/03/11/dining/induction-cooking.html.

———. "Mayonnaise: Oil, Egg and a Drop of Magic." *New York Times*, May 22, 2012, sec. Food. https://www.nytimes.com/2012/05/23/dining/easy-homemade-mayonnaise.html.

Cleveland Clinic. "9 Vitamins and Minerals You Should Take Daily." November 4, 2021. https://health.clevelandclinic.org/which-vitamins-should-you-take/.

CMI Orchards. "I'm Perfect." Accessed July 1, 2023. https://www.cmiapples.com/catalog/brands/im-perfect.

Condé Nast. "Why Are There 100 Folds in a Chef's Toque?" *Bon Appétit*, April 1, 2012. https://www.bonappetit.com/people/chefs/article/why-are-there-100-folds-in-a-chef-s-toque.

Cordell, Lyndsay. "The Reason Wagyu Beef Prices Are More Expensive Than Angus." April 4, 2022. https://www.wideopencountry.com/wagyu-beef-price/.

Cornell University, Department of Food Science. "Biomaterials and Biointerfaces." *Goddard Research Group* (blog). Accessed July 1, 2023. https://blogs.cornell.edu/goddard/.

Corteva Agriscience. "Welcome to Corteva Agriscience." Accessed July 1, 2023. https://www.corteva.com.

"Cultured Meat." Wikipedia. Assessed September, 26, 2024. https://en.wikipedia.org/wiki/Cultured_meat.

Dakin Farm. "Maple Sugar on Snow." Accessed July 1, 2023. https://www.dakinfarm.com/recipe.aspx/confection/Maple-Sugar-on-Snow.

Davoodi, Seyed Hossein, Roghiyeh Shahbazi, Saeideh Esmaeili, Sara Sohrabvandi, AmirMohamamd Mortazavian, Sahar Jazayeri, and Aghdas Taslimi. "Health-Related Aspects of Milk Proteins." *Iranian Journal of Pharmaceutical Research: IJPR* 15, no. 3 (2016): 573–91.

Denchak, Melissa. "Paris Climate Agreement: Everything You Need to Know." Natural Resources Defense Council, February 19, 2021. https://www.nrdc.org/stories/paris-climate-agreement-everything-you-need-know.

"Denis Papin." In *Encyclopedia Britannica*. Accessed August 18, 2024. https://www.britannica.com/biography/Denis-Papin.

De Oliveira, Diogo N., Maico De Menezes, and Rodrigo R. Catharino. "Thermal Degradation of Sucralose: A Combination of Analytical Methods to Determine Stability and Chlorinated Byproducts." *Scientific Reports* 5, no. 1 (April 15, 2015): 9598. https://doi.org/10.1038/srep09598.

Deveau, Hélène, Josiane E. Garneau, and Sylvain Moineau. "CRISPR/Cas System and Its Role in Phage-Bacteria Interactions." *Annual Review of Microbiology* 64, no. 1 (October 13, 2010): 475–93. https://doi.org/10.1146/annurev.micro.112408.134123.

Dorsey, Jenny. "Infrared Cooking 101." Institute of Culinary Education. January 22, 2021. https://www.ice.edu/blog/infrared-cooking.

Dubois, G. E., D. E. Walters, S. S. Schiffman, Z. S. Warwick, B. J. Booth, S. D. Pecore, K. Gibes, B. T. Carr, and L. M. Brans. "Concentration-Response Relationships of Sweeteners: A Systematic Study," In *Sweeteners: Discovery, Molecular Design and Chemoreception*, ed. D. E. Walters, F. T. Orthoefer, and G. E. Dubois. Vol. 450. ACS Symposium Series. Washington, DC: American Chemical Society, 1991. https://doi.org/10.1021/bk-1991-0450.

Egg Nutrition Facts. "Egg-Yolk101–187x300–1.Jpg (290×428)." Accessed July 1, 2023. https://eggs.ab.ca/wp-content/uploads/2021/03/Egg-Yolk101–187x300–1.jpg.

Erin. "How to Sous Vide Asparagus (with 3 Flavoring Ideas!)." Platings + Pairings, September 2022. https://www.platingsandpairings.com/sous-vide-asparagus/#recipe.

Everts, Sarah. "The Maillard Reaction Turns 100." *Chemical & Engineering News*, October 1, 2012. https://cen.acs.org/articles/90/i40/Maillard-Reaction-Turns-100.html.

Febbraio, Mark A., and Michael Karin. "'Sweet Death': Fructose as a Metabolic Toxin That Targets the Gut-Liver Axis." *Cell Metabolism* 33, no. 12 (December 2021): 2316–28. https://doi.org/10.1016/j.cmet.2021.09.004.

Fleming, Amy. "How Sound Affects the Taste of Our Food." *The Guardian*, March 11, 2014, sec. Life and style. https://www.theguardian.com/lifeandstyle/wordofmouth/2014/mar/11/sound-affects-taste-food-sweet-bitter.

"Foodpairing®—Better Products, Faster." July 24, 2021. https://www.foodpairing.com/.

Fox, A. L. "The Relationship between Chemical Constitution and Taste." *Proceedings of the National Academy of Sciences of the United States of America* 18, no. 1 (January 1932): 115–20. https://doi.org/10.1073/pnas.18.1.115.

Gibbs, W. Wayt, and Nathan Myhrvold. "Cryogenic Cooking." *Scientific American* 305, no. 2 (July 19, 2011): 31–31. https://doi.org/10.1038/scientificamerican0811-31.

Godoy, Maria. "Wal-Mart, America's Largest Grocer, Is Now Selling Ugly Fruit and Vegetables." NPR, July 20, 2016, sec. Producers. https://www.npr.org/sections/thesalt/2016/07/20/486664266/walmart-world-s-largest-grocer-is-now-selling-ugly-fruit-and-veg.

Goffeau, A., B. G. Barrell, H. Bussey, R. W. Davis, B. Dujon, H. Feldmann, F. Galibert, et al. "Life with 6000 Genes." *Science* 274, no. 5287 (October 25, 1996): 546–67. https://doi.org/10.1126/science.274.5287.546.

Good Food Institute. "The Science of Cultivated Meat." January 27, 2021. https://gfi.org/science/the-science-of-cultivated-meat/.

Gravina, Stephen A., Gregory L. Yep, and Mehmood Khan. "Human Biology of Taste." *Annals of Saudi Medicine* 33, no. 3 (May 2013): 217–22. https://doi.org/10.5144/0256–4947.2013.217.

Guan, Xin, Jingwen Zhou, Guocheng Du, and Jian Chen. "Bioprocessing Technology of Muscle Stem Cells: Implications for Cultured Meat." *Trends in Biotechnology* 40, no. 6 (June 2022): 721–34. https://doi.org/10.1016/j.tibtech.2021.11.004.

Hickenbottom, Sabrina J., Alan R. Kimmel, Constantine Londos, and James H Hurley. "Structure of a Lipid Droplet Protein." *Structure* 12, no. 7 (July 2004): 1199–1207. https://doi.org/10.1016/j.str.2004.04.021.

Hodge, J. E. "Dehydrated Foods, Chemistry of Browning Reactions in Model Systems." *Journal of Agricultural and Food Chemistry* 1, no. 15 (October 1953): 928–43. https://doi.org/10.1021/jf60015a004.

Holbrook, Eric H., and Donald A. Leopold. "An Updated Review of Clinical Olfaction." *Current Opinion in Otolaryngology & Head & Neck Surgery* 14, no. 1 (February 2006): 23–28. https://doi.org/10.1097/01.moo.0000193174.77321.39.

"Ice Cream Cone." Wikipedia. Accessed October 1, 2024. https://en.wikipedia.org/wiki/Ice_cream_cone.

Imdad, Alvi. "Difference between Conduction, Convection and Radiation in Tabular Form." *Ox Science* (blog). August 14, 2023. https://oxscience.com/conductionconvection-and-radiation/.

Inari. "Seeding Change to Bring Value across the Food System to Unlock Radical Possibilities." Accessed July 1, 2023. https://inari.com/.

"Induction Stoves and EMF Testing." California Energy Commission. September 25, 2020. https://efiling.energy.ca.gov/GetDocument.aspx?tn=234914&DocumentContentId=67776.

iSi Gourmet Whip How-to-Use. 2015. https://www.youtube.com/watch?v=uK2T9WxjUvc.

Jaglo, Kirsten, Shannon Kenny, and Jenny Stephenson. "From Farm to Kitchen: The Environmental Impacts of U.S. Food Waste." U.S. Environmental Protection Agency Office of Research and Development, November 2021. https://www.epa.gov/system/files/documents/2021-11/from-farm-to-kitchen-the-environmental-impacts-of-u.s.-food-waste_508-tagged.pdf.

Jeruzal-Świątecka, Joanna, Wojciech Fendler, and Wioletta Pietruszewska. "Clinical Role of Extraoral Bitter Taste Receptors." *International Journal of Molecular Sciences* 21, no. 14 (July 21, 2020): 5156. https://doi.org/10.3390/ijms21145156.

Jiang, Xiaoyi, Elhamalsadat Shekarforoush, Musemma Kedir Muhammed, Kathryn A. Whitehead, Nils Arneborg, and Jens Risbo. "Lactic Acid Bacteria as Structural Building Blocks in Non-Fat Whipping Cream Analogues." *Food Hydrocolloids* 135 (February 1, 2023): 108137. https://doi.org/10.1016/j.foodhyd.2022.108137.

"The Jetsons." Wikipedia. Accessed October 1, 2024. https://en.wikipedia.org/wiki/The_Jetsons.

Julius, David and Ardem Patapoutian. "Discoveries of Receptors for Temperature and Touch." Accessed July 1, 2023. https://www.nobelprize.org/prizes/medicine/2021/advanced-information/.

Julius, David. "TRP Channels and Pain." *Annual Review of Cell and Developmental Biology* 29, no. 1 (October 6, 2013): 355–84. https://doi.org/10.1146/annurev-cellbio-101011–155833.

Kanwal, Jessleen K. "Brain Tricks to Make Food Taste Sweeter: How to Transform Taste Perception and Why It Matters." *Science in the News* (blog), January 11, 2016. https://sitn.hms.harvard.edu/flash/2016/brain-tricks-to-make-food-taste-sweeter-how-to-transform-taste-perception-and-why-it-matters/.

Katidi, Alexandra, Konstantina Xypolitaki, Antonis Vlassopoulos, and Maria Kapsokefalou. "Nutritional Quality of Plant-Based Meat and Dairy Imitation Products and Comparison with Animal-Based Counterparts." *Nutrients* 15, no. 2 (January 12, 2023): 401. https://doi.org/10.3390/nu15020401.

Kim, JuJu. "Woman Loses Stomach after Drinking Liquid Nitrogen Cocktail." *Time*, October 9, 2012. https://newsfeed.time.com/2012/10/09/woman-loses-stomach-after-drinking-liquid-nitrogen-cocktail/.

Kitchen Theory. "Lecithin & Culinary Foam." *Immersive Dining* (blog). Accessed February 15, 2023. https://kitchen-theory.com/lecithin-culinary-foam/.

Klurfeld, D. M., J. Foreyt, T. J. Angelopoulos, and J. M. Rippe. "Lack of Evidence for High Fructose Corn Syrup as the Cause of the Obesity Epidemic." *International Journal of Obesity* 37, no. 6 (June 2013): 771–73. https://doi.org/10.1038/ijo.2012.157.

Kokaji, Nao, and Masashi Nakatani. "With a Hint of Sudachi: Food Plating Can Facilitate the Fondness of Food." *Frontiers in Psychology* 12 (October 15, 2021): 699218. https://doi.org/10.3389/fpsyg.2021.699218.

Kommenda, Niko, Naema Ahmed, Scott Dance, and Simon Ducroquet. "This Is the Environmental Cost of the Food We Eat." *Washington Post*, February 1, 2023. https://www.washingtonpost.com/climate-environment/interactive/2023/food-impact-climate-water-wildlife/.

La Cocina de Gisele. "Can You Buy 100 Year Old Eggs?" Accessed July 1, 2023. https://lacocinadegisele.com/knowledgebase/can-you-buy-100-year-old-eggs.

Lima, Miguel, Rui Costa, Ivo Rodrigues, Jorge Lameiras, and Goreti Botelho. "A Narrative Review of Alternative Protein Sources: Highlights on Meat, Fish, Egg and Dairy Analogues." *Foods* 11, no. 14 (July 11, 2022): 2053. https://doi.org/10.3390/foods11142053.

Liu, Dengyong, Yajun Deng, Lei Sha, Md. Abul Hashem, and Shengmei Gai. "Impact of Oral Processing on Texture Attributes and Taste Perception." *Journal of Food Science and Technology* 54, no. 8 (July 2017): 2585–93. https://doi.org/10.1007/s13197–017–2661–1.

Logsdon, Jason. "Modernist Cooking Made Easy." 100vampirenovels.net. Accessed December 11, 2023. https://100vampirenovels.net/pdf-novels/modernist-cooking-made-easy-by-jason-logsdon-free/8-page.

Ludwig, David S., Louis J. Aronne, Arne Astrup, Rafael De Cabo, Lewis C. Cantley, Mark I. Friedman, Steven B. Heymsfield, et al. "The Carbohydrate-Insulin Model: A Physiological Perspective on the Obesity Pandemic." *American Journal of Clinical Nutrition* 114, no. 6 (December 2021): 1873–85. https://doi.org/10.1093/ajcn/nqab270.

Ludwig, David S., and Cara B. Ebbeling. "The Carbohydrate-Insulin Model of Obesity: Beyond 'Calories In, Calories Out.'" *JAMA Internal Medicine* 178, no. 8 (August 1, 2018): 1098. https://doi.org/10.1001/jamainternmed.2018.2933.

Lunestad, Bjørn Tore, Didrik Hjertaker Grevskott, Irja Sunde Roiha, and Cecilie Smith Svanevik. "Microbiota of Lutefisk, a Nordic Traditional Cod Dish with a High pH." *Food Control* 90 (August 1, 2018): 312–16. https://doi.org/10.1016/j.foodcont.2018.03.011.

Madalene. "How to Make Culinary Foams, Air and Espumas." British Larder, April 2, 2015. https://www.britishlarder.co.uk/how-to-make-culinary-foams-air-and-espumas/.

Madju. "What Is the Difference between Stabilizers and Emulsifiers." Accessed December 6, 2022. https://www.differencebetween.com/what-is-the-difference-between-stabilizers-and-emulsifiers/.

Maicas, Sergi. "The Role of Yeasts in Fermentation Processes." *Microorganisms* 8, no. 8 (July 28, 2020): 1142. https://doi.org/10.3390/microorganisms8081142.

"Malted Milk." Wikipedia. Accessed October 1, 2024. https://en.wikipedia.org/wiki/Malted_milk.

MasterClass. "Century Eggs Explained: How to Make Century Eggs—2023." December 4, 2023. https://www.masterclass.com/articles/century-eggs-explained-plus-century-egg-recipe.

McClements, David Julian, and Lutz Grossmann. "A Brief Review of the Science behind the Design of Healthy and Sustainable Plant-Based Foods." *Npj Science of Food* 5, no. 1 (June 3, 2021): 17. https://doi.org/10.1038/s41538-021-00099-y.

McClements, David Julian, Emily Newman, and Isobelle Farrell McClements. "Plant-Based Milks: A Review of the Science Underpinning Their Design, Fabrication, and Performance." *Comprehensive Reviews in Food Science and Food Safety* 18, no. 6 (November 2019): 2047–67. https://doi.org/10.1111/1541-4337.12505.

McGee, Harold. "Caramelization: New Science, New Possibilities." Curious Cook. Accessed July 1, 2023. https://www.curiouscook.com/site/2012/09/caramelization-new-science-new-possibilities.html.

———. "For Old-Fashioned Flavor, Bake the Baking Soda." *New York Times*, September 14, 2010, sec. Food. https://www.nytimes.com/2010/09/15/dining/15curious.html.

———. *On Food and Cooking: The Science and Lore of the Kitchen.* Completely rev. and updated. New York: Scribner, 2004.

McGovern, Patrick, Mindia Jalabadze, Stephen Batiuk, Michael P. Callahan, Karen E. Smith, Gretchen R. Hall, Eliso Kvavadze, et al. "Early Neolithic Wine of Georgia in the South Caucasus." *Proceedings of the National Academy of Sciences* 114, no. 48 (November 28, 2017). https://doi.org/10.1073/pnas.1714728114.

Meyer, Abby, and Sara Dastgheib-Vinarov. "The Future of Food? CRISPR-Edited Agriculture." *Food and Drug Law Institute (FDLI)* (blog). November 19, 2021. https://www.fdli.org/2021/11/the-future-of-food-crispr-edited-agriculture/.

"Microwave Ovens." Accessed July 1, 2023. http://hyperphysics.phy-astr.gsu.edu/hbase/Waves/mwoven.html.

Misfits Market. "Grocery Delivery FAQs." Accessed July 1, 2023. https://www.misfitsmarket.com/faq.

Moayedi, Yalda, Lucia F. Duenas-Bianchi, and Ellen A. Lumpkin. "Somatosensory Innervation of the Oral Mucosa of Adult and Aging Mice." *Scientific Reports* 8, no. 1 (July 2, 2018): 9975. https://doi.org/10.1038/s41598-018-28195-2.

Molecular Recipes. "Strawberry Spheres with Ground Pepper." December 12, 2013. http://www.molecularrecipes.com/spherification-1/strawberry-spheres-ground-pepper/.

Morone, Manny "Fox." "Forget Expiration Dates. Spoilage Sensors Could Tell Us When Food Actually Goes Bad." *Chemical & Engineering News*, May 17, 2020. https://cen.acs.org/analytical-chemistry/diagnostics/Forget-expiration-dates-Spoilage-sensors/98/i19.

Moskin, Julia, Brad Plumer, Rebecca Lieberman, Eden Weingart, and Nadja Popovich. "Your Questions about Food and Climate Change, Answered." *New York Times*, April 30, 2019, sec. Food. https://www.nytimes.com/interactive/2022/dining/climate-change-food-eating-habits.html, https://www.nytimes.com/interactive/2022/dining/climate-change-food-eating-habits.html.

Myhrvold, Nathan. *Modernist Cuisine.* Vol. 4: *The Art and Science of Cooking.* Seattle: Cooking Lab, 2011.

Naddaf, Miryam. "Aspartame Is a Possible Carcinogen: The Science behind the Decision." *Nature*, July 14, 2023, d41586-023-02306-0. https://doi.org/10.1038/d41586-023-02306-0.

National Cancer Institute. "Definition of Homeostasis—NCI Dictionary of Cancer Terms—NCI." nciAppModulePage, February 2, 2011. Nciglobal, ncienterprise. https://www.cancer.gov/publications/dictionaries/cancer-terms/def/homeostasis.

National Institutes of Health (NIH). "Gene Editing—Digital Media Kit." March 7, 2019. https://www.nih.gov/news-events/gene-editing-digital-press-kit.

National Restaurant Association Educational Foundation. "ServSafe—Get Certified at Home." Accessed July 1, 2023. http://servsafe.com/Home.

NobelPrize.org. "The Nobel Prize in Chemistry 2020." Accessed July 1, 2023. https://www.nobelprize.org/prizes/chemistry/2020/summary/.

Nolden, Alissa A., and John E. Hayes. "Perceptual and Affective Responses to Sampled Capsaicin Differ by Reported Intake." *Food Quality and Preference* 55 (January 2017): 26–34. https://doi.org/10.1016/j.foodqual.2016.08.003.

Nolden, Alissa A., Gabrielle Lenart, and John E. Hayes. "Putting out the Fire—Efficacy of Common Beverages in Reducing Oral Burn from Capsaicin." *Physiology & Behavior* 208 (September 2019): 112557. https://doi.org/10.1016/j.physbeh.2019.05.018.

Novotny, Janet A., Sarah K. Gebauer, and David J. Baer. "Discrepancy between the Atwater Factor Predicted and Empirically Measured Energy Values of Almonds in Human Diets." *American Journal of Clinical Nutrition* 96, no. 2 (August 2012): 296–301. https://doi.org/10.3945/ajcn.112.035782.

O'Brian, Mark R.. "How Does the Impossible Burger Look and Taste Like Real Beef?" *Discover Magazine*, April 22, 2019. https://www.discovermagazine.com/planet-earth/how-does-the-impossible-burger-look-and-taste-like-real-beef.

OneGenAway. "One Generation Away." Accessed August 11, 2022. https://onegenaway.com/.

Palm, Gottfried J., Lukas Reisky, Dominique Böttcher, Henrik Müller, Emil A. P. Michels, Miriam C. Walczak, Leona Berndt, Manfred S. Weiss, Uwe T. Bornscheuer, and Gert Weber. "Structure of the Plastic-Degrading Ideonella Sakaiensis MHETase Bound to a Substrate." *Nature Communications* 10, no. 1 (April 12, 2019): 1717. https://doi.org/10.1038/s41467-019-09326-3.

Patel, Prachi. "The Time Is Now for Edible Packaging." *Chemical & Engineering News*, January 26, 2020. https://cen.acs.org/food/food-science/time-edible-packaging/98/i4.

Poore, J., and T. Nemecek. "Reducing Food's Environmental Impacts through Producers and Consumers." *Science* 360, no. 6392

(June 2018): 987–92. https://doi.org/10.1126/science.aaq0216.

Popovich, Nadja, Cari Vander Yacht, and Eden Weingart. "Quiz: How Does Your Diet Contribute to Climate Change?" *New York Times*, April 30, 2019, sec. Climate. https://www.nytimes.com/interactive/2019/04/30/climate/your-diet-quiz-global-warming.html, https://www.nytimes.com/interactive/2019/04/30/climate/your-diet-quiz-global-warming.html.

Popovici, Alice. "The Colonel in the Kitchen: A Surprising History of Sous Vide." NPR, May 9, 2018, sec. Food History and Culture. https://www.npr.org/sections/thesalt/2018/05/09/608308624/the-colonel-in-the-kitchen-a-surprising-history-of-sous-vide.

Population Connection. "Population Milestones." Accessed July 1, 2023. https://populationconnection.org/learn/population-milestones/.

Project Drawdown. "Drawdown Foundations." February 5, 2020. https://drawdown.org/drawdown-foundations.

Rennie, Helen. "Random Beets and How to Make Vegetable Foams." *Beyond Salmon* (blog). July 29, 2012. http://www.beyondsalmon.com/2012/07/random-beets-and-how-to-make-vegetable.html.

Rodriguez, Douglas, and Laura Zimmerman. *The Great Ceviche Book*. Berkeley, CA: Ten Speed Press, 2003.

Romani, Viviane P., Vilásia G. Martins, and Julie M. Goddard. "Radical Scavenging Polyethylene Films as Antioxidant Active Packaging Materials." *Food Control* 109 (March 2020): 106946. https://doi.org/10.1016/j.foodcont.2019.106946.

Roper, Stephen D. "TRPs in Taste and Chemesthesis." In *Mammalian Transient Receptor Potential (TRP) Cation Channels*, ed. Bernd Nilius and Veit Flockerzi, 223:827–71. Handbook of Experimental Pharmacology. Cham: Springer, 2014. https://doi.org/10.1007/978–3-319–05161–1_5.

Roy, Sudhir C. "Low-Density Lipoproteins from Egg Yolk: A Natural Carrier of Highly Emulsifying Species." *New Food Magazine*, February 28, 2013. https://www.newfoodmagazine.com/article/10090/low-density-lipoproteins-from-egg-yolk-a-natural-carrier-of-highly-emulsifying-species/.

Ryp Labs. "Fighting Food Waste." Accessed July 1, 2023. https://www.ryplabs.com.

Şahin, Esra. "A Bibliometric Overview of the International Journal of Gastronomy and Food Science: To Where Is Gastronomy Research Evolving?" *International Journal of Gastronomy and Food Science* 28 (June 2022): 100543. https://doi.org/10.1016/j.ijgfs.2022.100543.

Sáinz, Neira, Carlos J. González-Navarro, J. Alfredo Martínez, and Maria J. Moreno-Aliaga. "Leptin Signaling as a Therapeutic Target of Obesity." *Expert Opinion on Therapeutic Targets* 19, no. 7 (July 3, 2015): 893–909. https://doi.org/10.1517/14728222.2015.1018824.

Sandman. "The Science Behind Caramelization." *The Food Untold* (blog). August 5, 2021. https://thefooduntold.com/food-chemistry/the-science-behind-caramelization/.

SBS Food. "The Science of Ceviche—Is It Cooked?" Accessed July 1, 2023. https://www.sbs.com.au/food/article/the-science-of-ceviche-is-it-cooked/jwi7c9ydh.

"Science of Eggs: Anatomy of an Egg." Exploratorium: Museum of Science, Art, and Human Perception. Accessed July 1, 2023. https://annex.exploratorium.edu/cooking/eggs/eggcomposition.html.

Scott, Alex. "Nestlé Will Spend $2 Billion to Peel Away from Virgin Plastic." *Chemical & Engineering News*, January 23, 2020. https://cen.acs.org/materials/polymers/Nestlspend-2-billion-peel-away/98/i4.

Scully, Crispian. *Oral and Maxillofacial Medicine: The Basis of Diagnosis and Treatment*. 3rd ed. Edinburgh: Churchill Livingstone/Elsevier, 2013.

Šimat, Vida, Nikheel Bhojraj Rathod, Martina Čagalj, Imen Hamed, and Ivana Generalić Mekinić. "Astaxanthin from Crustaceans and Their Byproducts: A Bioactive Metabolite Candidate for Therapeutic Application." *Marine Drugs* 20, no. 3 (March 12, 2022): 206. https://doi.org/10.3390/md20030206.

Smithsonian Magazine and K. Annabelle Smith. "Why the Tomato Was Feared in Europe for More Than 200 Years." *Smithsonian Magazine*. Accessed July 1, 2023. https://www.smithsonianmag.com/arts-culture/why-the-tomato-was-feared-in-europe-for-more-than-200-years-863735/.

Souza, Dan. "Perfect Scrambled Eggs." America's Test Kitchen. Accessed July 1, 2023. https://www.americastestkitchen.com/cooksillustrated/videos/2003-perfect-scrambled-eggs?incode=MASAZ00L0&ref=search_results_1.

"Speyer Wine Bottle." Wikipedia. Accessed October 1, 2024. https://en.wikipedia.org/wiki/Speyer_wine_bottle.

Starowicz, Małgorzata. "Analysis of Volatiles in Food Products." *Separations* 8, no. 9 (September 17, 2021): 157. https://doi.org/10.3390/separations8090157.

Statista. "U.S. Population: Most Eaten Brands of Whipped Topping (Cream Type) 2011–2020." Accessed July 1, 2023. https://www.statista.com/statistics/281278/us-households-most-eaten-brands-of-whipped-topping-cream-type-trend/.

Stevens Institute of Technology. "Direct Touch of Food Makes Eating Experience More Enjoyable." ScienceDaily, February 5, 2020. https://www.sciencedaily.com/releases/2020/02/200205151511.htm.

Sugar: The Bitter Truth. YouTube video, 2009. https://www.youtube.com/watch?v=dBnniua6-oM.

Sullivan, Kevin P., Allison Z. Werner, Kelsey J. Ramirez, Lucas D. Ellis, Jeremy R. Bussard, Brenna A. Black, David G. Brandner, et al. "Mixed Plastics Waste Valorization through Tandem Chemical Oxidation and Biological Funneling." *Science* 378, no. 6616 (October 14, 2022): 207–11. https://doi.org/10.1126/science.abo4626.

Sun, Chen, and Shicui Zhang. "Immune-Relevant and Antioxidant Activities of Vitellogenin and Yolk Proteins in Fish." *Nutrients* 7, no. 10 (October 22, 2015): 8818–29. https://doi.org/10.3390/nu7105432.

Tamanna, Nahid, and Niaz Mahmood. "Food Processing and Maillard Reaction Products: Effect on Human Health and Nutrition." *International Journal of Food Science* 2015 (2015): 1–6. https://doi.org/10.1155/2015/526762.

Towell, Colin. *Essential Survival Skills*. London: Penguin, 2011. https://www.dk.com/us/book/9780756659981-essential-survival-skills/.

TreeCo. "R&D Engine of the Forest Industry." Accessed July 1, 2023. https://tree-co.com/.

United Nations. "Day of 8 Billion." Accessed July 1, 2023. https://www.un.org/en/dayof8billion.

———. "The Paris Agreement." Accessed July 1, 2023. https://www.un.org/en/climatechange/paris-agreement.

———. "What Is Climate Change?" Accessed July 1, 2023. https://www.un.org/en/climatechange/what-is-climate-change.

U.S. Congress. Nutrition Labeling and Education Act of 1990. Pub. L. No. 535, U.S. Statutes at Large, vol. 104, 101st Cong., 2nd sess. (1990). https://www.govinfo.gov/app/details/STATUTE-104/STATUTE-104-Pg2353.

———. House of Representatives. "Food Labeling Modernization Act of 2023." H.R. 2901. Pub. L. No. 2901. 118th Cong. (2023). https://www.congress.gov/bill/118th-congress/house-bill/2901.

U.S. Department of Agriculture (USDA). "High Altitude Cooking." June 13, 2015. http://www.fsis.usda.gov/food-safety/safe-food-handling-and-preparation/food-safety-basics/high-altitude-cooking.

———. "Is It Safe to Eat Cheese Made from Unpasteurized Milk?" March 6, 2023. https://ask.usda.gov/s/article/Is-it-safe-to-eat-cheese-made-from-unpasteurized-milk.

U.S. Food and Drug Administration (FDA). "Food Code 2022." November 6, 2023. https://www.fda.gov/food/fda-food-code/food-code-2022.

U.S. Environmental Protection Agency (EPA). "Wasted Food Scale." Overviews and Factsheets. August 12, 2015. https://www.epa.gov/sustainable-management-food/wasted-food-scale.

Venkatachalam, Kartik, and Craig Montell. "TRP Channels." *Annual Review of Biochemistry* 76, no. 1 (June 7, 2007): 387–417. https://doi.org/10.1146/annurev.biochem.75.103004.142819.

Vines, Brian. "Ugly Food Fight: Misfits Market, Imperfect Foods, and the Battle against Food Waste." *Consumer Reports*, January 12, 2022. https://www.consumerreports.org/health/food-shopping/ugly-food-fight-misfits-market-imperfect-foods-food-waste-a6326488257/.

Walters, D. Eric, Frank T. Orthoefer, and Grant E. DuBois, eds. *Sweeteners: Discovery, Molecular Design, and Chemoreception*. Vol. 450. ACS Symposium Series. Washington, DC: American Chemical Society, 1991. https://doi.org/10.1021/bk-1991-0450.

Wei, Shaobo, Xia Li, Zefu Lu, Hui Zhang, Xiangyuan Ye, Yujie Zhou, Jing Li, et al. "A Transcriptional Regulator That Boosts Grain Yields and Shortens the Growth Duration of Rice." *Science* 377, no. 6604 (July 22, 2022): eabi8455. https://doi.org/10.1126/science.abi8455.

What-When-How.com. "Tmp4697_thumb1.Jpg (965×768)." Accessed July 1, 2023. http://what-when-how.com/wp-content/uploads/2011/05/tmp4697_thumb1.jpg.

Whenish, Ruban, Seeram Ramakrishna, Amit Kumar Jaiswal, and Geetha Manivasagam. "A Framework for the Sustainability Implications of 3D Bioprinting through Nature-Inspired Materials and Structures." *Bio-Design and Manufacturing* 5, no. 2 (April 2022): 412–23. https://doi.org/10.1007/s42242-021-00168-x.

White House. "Biden-Harris Administration National Strategy on Hunger, Nutrition, and Health." September 2022. https://www.whitehouse.gov/wp-content/uploads/2022/09/White-House-National-Strategy-on-Hunger-Nutrition-and-Health-FINAL.pdf.

Williams Sonoma. "Rösle Egg Topper." Accessed July 1, 2023. https://www.williams-sonoma.com/products/rosle-egg-topper/.

Woodford, Chris. "How Do Induction Cooktops Work?" Explain That Stuff, April 8, 2011. http://www.explainthatstuff.com/induction-cooktops.html.

Wooding, Stephen. "Phenylthiocarbamide: A 75-Year Adventure in Genetics and Natural Selection." *Genetics* 172, no. 4 (April 1, 2006): 2015–23. https://doi.org/10.1093/genetics/172.4.2015.

World Central Kitchen. "World Central Kitchen." Accessed July 1, 2023. https://wck.org/.

———. "WCK Serves 12+ Million Meals in the Month since Türkiye-Syria Quakes." March 10, 2023. https://wck.org/news/turkiye-syria-update2.

World Health Organization (WHO). "Obesity and Overweight." Accessed July 1, 2023. https://www.who.int/news-room/fact-sheets/detail/obesity-and-overweight.

XiaoZhi Lim. "Xiaogang Liu Is Using Fluorescence to Tackle the Problem of Illicit Cooking Oil." *Chemical & Engineering News*, May 2, 2020. https://cen.acs.org/food/food-science/Xiaogang-Liu-using-fluorescence-tackle/98/i17.

Yang, Chun Z., Stuart I. Yaniger, V. Craig Jordan, Daniel J. Klein, and George D. Bittner. "Most Plastic Products Release Estrogenic Chemicals: A Potential Health Problem That Can Be Solved." *Environmental Health Perspectives* 119, no. 7 (July 1, 2011): 989–96. https://doi.org/10.1289/ehp.1003220.

Yeung, Marie. "Microbial Forensics in Food Safety." *Microbiology Spectrum* 4, no. 4 (August 12, 2016): 10.1128/microbiolspec.emf-0002–2013. https://doi.org/10.1128/microbiolspec.emf-0002–2013.

Zaidi, Syed Shan-e-Ali, Hervé Vanderschuren, Matin Qaim, Magdy M. Mahfouz, Ajay Kohli, Shahid Mansoor, and Mark Tester. "New Plant Breeding Technologies for Food Security." *Science* 363, no. 6434 (March 29, 2019): 1390–91. https://doi.org/10.1126/science.aav6316.

Zampini, Massimiliano, and Charles Spence. "The Role of Auditory Cues in Modulating the Perceived Crispness and Staleness of Potato Chips." *Journal of Sensory Studies* 19, no. 5 (October 2004): 347–63. https://doi.org/10.1111/j.1745-459x.2004.080403.x.

Zimmermann, Lisa, Georg Dierkes, Thomas A. Ternes, Carolin Völker, and Martin Wagner. "Benchmarking the In Vitro Toxicity and Chemical Composition of Plastic Consumer Products." *Environmental Science & Technology* 53, no. 19 (October 1, 2019): 11467–77. https://doi.org/10.1021/acs.est.9b02293.

Illustration Credits

All molecular structures, graphs, and cartoons are by Richard A. Blatchly unless otherwise noted.

Figure 2.1 Fernando Da Cunha / Science Photo Library C049/3364. https://www.sciencephoto.com/media/1120264/view.

Figure 2.2 Illustration by Lydia V. Kibiuk, Baltimore, MD and Devon Stuart, Harrisburg, PA, used with permission from The Society for Neuroscience.

Figure 2.4 AdobeStock 374884753 Educational License

Figure 2.5 Copyright © 2010 Chaudhari and Roper

Figure 2.7 S. D. Roper, (2014). "TRPs in Taste and Chemesthesis," in *Mammalian Transient Receptor Potential (TRP) Cation Channels*, ed. B. Nillus and V. Flockerzi, Handbook of Experimental Pharmacology, vol. 223 (Cham: Springer, 2014), https://doi.org/10.1007/978–3-319–05161–1_5. License 5706100587641.

Figure 2.8 Illustration by Efazzari.

Figure 2.10 Data from James Briscione and Brooke Parkhurst, *The Flavor Matrix: The Art and Science of Pairing Common Ingredients to Create Extraordinary Dishes* (New York: Houghton Mifflin Harcourt, 2018).

Figure 3.1 Used with permission from NASA. Illustration in https://www.machinedesign.com/learning-resources/whats-the-difference-between/document/21834474/whats-the-difference-between-conduction-convection-and-radiation

Figure 3.3 Image by GSTraub. EmSpectrum_iStock-1271622591

Figure 3.7 Adapted from https://www.jonbarron.org/uncategorized/healthiest-cooking-oil-comparison-chart-with-smoke-points-and-omega-3-fatty-acid-ratios/ and Wikipedia http://en.wikipedia.org/wiki/Smoke_point

Figure 3.11 iStock-526312443

Figure 3.12 From *The Role of Proteins in Foods—Cooking and Denaturation* and shared under a CC BY-NC-SA 3.0 license and was authored, remixed, and/or curated by LibreTexts

Figure 3.13 Used with permission from Dave Arnold, https://cookingissues.com/wp-content/uploads/2011/04/The_Egg_Chart.jpg.

Figure 4.2 Copyright © 2002 Richard Megna—Fundamental Photographs

Figure 4.4 Ceviche by Picanteria karol CC BY-SA 4.0. https://commons.wikimedia.org/w/index.php?curid=34894146. Lutefisk: CC BY-SA 3.0 2023 https://en.wikipedia.org/wiki/Lutefisk Lutefisk. Century Egg (Shutterstock Image _83871829(1).

Figure 4.6 Jan Christoph Peinemann, Chaeyoung Rhee, Seung Gu Shin, Daniel Pleissner, "Non-Sterile Fermentation of Food Waste with Indigenous Consortium and Yeast—Effects on Microbial Community and Product Spectrum," *Bioresource Technology* 306 (2020): 123175. Elsevier License 5756220297765 (food waste)

Figure 5.1 Photos by Richard A. Blatchly

Figure 5.3 Cartoon created by Donna Nestor

Figure 5.6 Images collected using differential interference contrast in collaboration with L. Panagis at the Amherst College Molecular Imaging Center.

Figure 5.7 Illustration by David S. Goodsell, RCSB Protein Data Bank. doi: 10.2210/rcsb_pdb/goodsell-gallery-039

Figure 5.8 D. J. McClements and L. Grossmann "Plant-Based Milk and Cream Analogs," in *Next-Generation Plant-Based Foods* (Cham: Springer, 2022), https://doi.org/10.1007/978–3–030–96764–2_8. License Number: 5470671695

Figure 6.1–6.2 Copyright © 2021, Project Drawdown. Used with permission from Project Drawdown, www.drawdown.org

Figure 6.3 Adapted from https://www.science.org/doi/full/10.1126/science.aaq0216

Figure 6.4 S. M. Andler and J. M. Goddard "Transforming food Waste: How Immobilized Enzymes Can Valorize Waste Streams into Revenue Streams," *npj Science of*

Food 2, no. 19 (2018), https://doi.org/10.1038/s41538–018–0028–2 Creative Commons CC BY 4.0

Figure 6.5 Viviane P. Romani, Vilásia G. Martins, and Julie M. Goddard, Radical Scavenging Polyethylene Films as Antioxidant Active Packaging Materials, *Food Control* 109 (2020):106946, ISSN 0956–7135, https://doi.org/10.1016/j.foodcont.2019.106946. License #: 5704370061110

Figure 6.6 K. P. Sullivan, A. Z. Werner, K. J. Ramirez, L. D. Ellis, J. R. Bussard, B. A. Black, D. G. Brandner, F. Bratti, B. L. Buss, X. Dong, S. J. Haugen, M. A. Ingraham, M. O. Konev, W. E. Michener, J. Miscall, I. Pardo, S. P. Woodworth, A. M. Guss, Y. Román-Leshkov, S. S. Stahl, and G. T. Beckham, "Mixed Plastics Waste Valorization through Tandem Chemical Oxidation and Biological Funneling, *Science* 378, no. 6616 (October 14, 2022): 207–11. doi: 10.1126/science.abo4626. Epub October 13, 2022. PMID: 36227984. License # 5704380854754. Copyright © 2022 the authors, Licensee American Association for the Advancement of Science. No claim to original US government works. https://www.science.org/about/science-licenses-journal-article-reuse

Figure 6.7 From National Institutes of Health, "Gene Editing—Digital Media Kit," https://www.nih.gov/news-events/gene-editing-digital-press-kit. Permission from NIH website

Figure 6.8 Wei, S., Li, X., Lu, Z., Zhang, H., Ye, X., Zhou, Y., Li, J., Yan, Y., Pei, H., Duan, F., Wang, D., Chen, S., Wang, P., Zhang, C., Shang, L., Zhou, Y., Yan, P., Zhao, M., Huang, J., . . . Zhou, W. (2022). A transcriptional regulator that boosts grain yields and shortens the growth duration of rice. *Science*. https://doi.org/abi8455 DOI: 10.1126/science.abi8455. License information: # 1435506–1? Copyright © 2022 the authors, some rights reserved; exclusive licensee American Association for the Advancement of Science. No claim to original US government works. https://www.science.org/about/science-licenses-journal-article-reuse

Figure 6.9 H. Zhang, F. Zhou, Y. Kan, X. Shan, W. Ye, Q. Dong, T. Guo, H. Xiang, B. Yang, C. Li, Y. Zhao, X. Yu, Q. Lu, Q. Guo, J. Lei, B. Liao, R. Mu, J. Cao, Y. Yu, . . . X. Lin, "A Genetic Module at One Locus in Rice Protects Chloroplasts to Enhance Thermotolerance," *Science* (2022), https://doi.org/10.1126/science.abo5721 License information: 1435513–1 Copyright © 2022 the Authors, some rights reserved, exclusive licensee American Association for the Advancement of Science. No claim to original US government works. https://www.science.org/about/science-licenses-journal-article-reuse

Figure 6.10 (a) Used with permission from AMSBIO https://www.amsbio.com/cultured-meat/ created using BioRender.com. (b) From Shutterstock, Stock Photo ID: 142419734, photo contributor: kitreel

Figure 6.11 https://www.creativemachineslab.com/laser-cooking.html Used with permission from H. Lipson at Columbia University.

Figure 6.12 Used with permission: Photographs by Hevré This. As appeared on *Nova*, https://www.pbs.org/wgbh/nova/article/synthetic-food-better-cooking-through-chemistry/

Figure 7.2 Photos by Brandon Kwon

Figure 7.3 iStock Image Stock illustration ID:1415661939 image credit Dmytro Omelianenko

Figure 7.4 H. Lu, J. A. Butler, N. S. Britten, P. D. Venkatraman, and S. S. Rahatekar, "Natural Antimicrobial Nano Composite Fibres Manufactured from a Combination of Alginate and Oregano Essential Oil," *Nanomaterials* 11 (8): 2062. https://doi.org/10.3390/nano11082062. CC BY 4.0

Figure 7.5 Photo by the author

Index

Note: Tables, figures, and boxes are indicated by *t*, *f*, or *b* after the page number.

www.ingramcontent.com/pod-product-compliance
Ingram Content Group UK Ltd.
Pitfield, Milton Keynes, MK11 3LW, UK
UKHW051833140825
461712UK00008B/58